SpringerBriefs in Electrical and Computer Engineering

For further volumes:
http://www.springer.com/series/10059

Rohit Sharma · Tapas Chakravarty

Compact Models
and Measurement Techniques
for High-Speed Interconnects

Rohit Sharma
Interconnect Focus Center
Georgia Institute of Technology
Atlanta, GA, USA

Tapas Chakravarty
Tata Consultancy Services
Kolkata, India

ISSN 2191-8112
ISBN 978-1-4614-1070-6
DOI 10.1007/978-1-4614-1071-3
Springer New York Heidelberg Dordrecht London

e-ISSN 2191-8120
e-ISBN 978-1-4614-1071-3

Library of Congress Control Number: 2012931862

Printed on acid-free paper

Springer is part of Springer Science+Business Media (www.springer.com)

There is something fascinating about science. One gets such wholesale returns of conjecture out of such a trifling investment of fact.

Mark Twain

Foreword

High-speed interconnect analysis and design has been the focus of multi-disciplinary research activities involving new materials, structures and components. Interconnect performance can be a major design bottleneck in overall system performance, as described by researchers for nearly three decades. Research has led to technological breakthroughs in interconnect design and structures which paralleled transistor scaling and enabled system growth. The Georgia Institute of Technology has been one of the institutions where interconnect and packaging research has been a focus. Georgia Tech is the home of the longest running, externally funded center of excellence in high-speed interconnect, namely the Interconnect Focus Center (IFC), one of six focus centers sponsored by the semiconductor industry and DARPA through the Semiconductor Research Corporation. The IFC undertakes pioneering research work in electrical, optical and carbon interconnects.

The design of chip-to-chip interconnect is an area which needs increased attention due to the challenges in off-chip bandwidth and energy consumption. High frequency signal losses and bandwidth requirements, as outlined by the projections in the International Technology Roadmap for Semiconductors, give a description of the simultaneous need to reduce interconnect dimension, increase the data rate, and lower the energy consumed. This set of goals present challenges for device fabrication and characterization, which require development of complex analytical models, time-intensive simulations and state-of-the-art measurements. However, there are few sources available in the literature that exclusively discuss chip-to-chip interconnect. To this end, I feel the present text offers valuable insights for readers interested in this area of research and development. The authors have put forward a unified analytical modeling approach specifically for the analysis of chip-to-chip planar interconnects. In my view, this is an important first step toward the analysis and optimization of complex interconnect layouts governed by material and performance constraints. The inclusion of measurement and simulation as a part of the text would provide a holistic understanding of the subject matter to the readers.

This book is the outcome of on-going research by the Georgia Tech authors and is supported by research funding, such as provided by the IFC. Researchers here and throughout the world will hopefully find this book instrumental in advancing product design and analysis. This book takes the reader from a simple introduction to this area of research to a complete understanding of the modeling and measurement techniques. I congratulate the authors for their efforts in bringing this book to fruition and for contributing to the IFC advances at Georgia Tech.

Atlanta

Paul A. Kohl
Interconnect Focus Center
311 Ferst Dr.
Atlanta, GA 30332
USA

Preface

High-speed interconnects are essentially wires that form the media for transmission of analog and digital signals in electronic circuits and systems. Along with devices, these interconnects form a dense and complex fabric that is responsible for performance of integrated circuits, boards and packages. Signal integrity in high-speed interconnects is one of the most important design aspects for achieving high performance and throughput. For most of the early *IC* era, designers and practicing engineers focused on device improvement alone. While computational capability of devices is important, overall system performance will hit a plateau level if the performance of interconnects is not improved. With increased clock speeds and reduced aspect ratios, interconnects became the most crucial design bottleneck. This led to unprecedented thrust on the design and analysis of interconnects, both in the semiconductor industry as well in academia. Over the years it has become a major research theme in the *ITRS* predictions as well as several industrial and academic journals and conferences.

This book will provide a detailed analysis of issues related to high-speed interconnects from the perspective of modeling approaches and measurement techniques. In that we restrict ourselves to electrical chip-chip interconnects. Particular focus is laid on the unified approach (variational method combined with the transverse transmission line technique) to develop efficient compact models for planar interconnects. This book will give a qualitative summary of the various reported modeling techniques and approaches and will help researchers and graduate students with deeper insights into interconnect models in particular and interconnect in general. Time domain and frequency domain measurement techniques and simulation methodology are also explained in this book.

The book is organized into four chapters. Chapter 1 discusses the evolution of interconnects as a research theme from a historical perspective. From the simplistic lumped *RC* regime to more complex transmission line models, interconnect modeling has truly come of age. The importance of high-speed effects and its relevance to signal integrity is covered in this chapter. A brief historical outlook on interconnects starting from the early *RC* era is provided. Brief overviews of the technological evolution of interconnect technology and its influence on modeling approaches is presented. However, for well-informed readers, having a glance at this introductory chapter should be sufficient. Chapter 2 explains the basics of compact model development for interconnects. We qualitatively summarize some of the most widely used analytical approaches toward analyzing transmission line interconnects. The chapter also presents the unified approach which is essentially a combination of variational analysis and transverse transmission line technique. The merit of this approach and its applicability is clearly explained to the readers. Application of the unified approach to develop compact physical models for high-speed interconnects is explained in Chap. 3. Models for parasitic extraction, computation of line impedance and time domain analysis of high-speed interconnects are explained. While the discussion is limited to a few useful interconnect structures it is felt that readers should be able to apply the technique to a wide variety of interconnect geometries. In Chap. 4, we discuss measurement techniques, wherein time domain and frequency domain measurement techniques are presented. Simulation methodology and numerical modeling approach for interconnects are also presented briefly.

This book will serve as a platform for the basic understanding of compact interconnect models using the unified approach. It also clearly explains measurement techniques and simulation methodologies for chip-chip interconnects to researchers and graduate students alike. As a note, this book is not a text book but will rather best fit as a reference book for students who are initiated to the area of interconnect modeling and measurements.

Rohit Sharma
Tapas Chakravarty

Acknowledgments

The area of interconnects in general and interconnect modeling in particular is a fascinating research theme. However, due to the concise nature of this book it was quite challenging to summarize vast amount of literature in a limited printed space. Also, the fact that the entire text had to completed in a short span made this task even more challenging. The authors thank the Almighty for His divine motivation to overcome several difficult phases during the preparation of this text.

The authors would also like to thank the Interconnect Focus Center at Georgia Institute of Technology and the Indo-US Science and Technology Forum. We thank all our friends and colleagues in Georgia Tech with whom we have had many insightful discussions. During the last several months the authors received exceptional support and cooperation from the editorial office of Springer, NY. In particular the authors thank Alex Greene and Allison Michael of the Springer Briefs series for their support and patience and appreciate their editorial work.

Finally, we acknowledge the love and care we received during these months from our families. It is only because of their perseverance and patience that we could complete this work in a timely manner. We dedicate this work to our family members.

Contents

Acronyms

ITRS	International technology roadmap for semiconductors
IC	Integrated circuits
CAD	Computer aided design
NoC	Network on chip
SoC	System on chip
TSV	Through silicon via
PCB	Printed circuit board
TEM	Transverse electromagnetic
EM	Electromagnetic
PEEC	Partial element equivalent circuit
MIC	Microwave integrated circuits
MMIC	Monolithic microwave integrated circuits
MCM	Multichip module
ISM	Industrial, scientific and medical
DVC	Discrete variational conformal
SPICE	Simulation program for integrated circuit environment
FDTD	Finite difference time domain
FEM	Finite element method
MoM	Method of moment
TDR	Time domain reflectometry
TDT	Time domain transmission
FDR	Frequency domain reflectometry
IFFT	Inverse fast fourier transform
TRL	Thru' reflect line
VNA	Vector network analyzer

Symbols

.

λ	Wavelength
G	Green's function
Y	Admittance parameter
Z, Z_0	Impedance parameter
C	Capacitance
Q	Charge
W_e	Electrostatic energy
V	Electric potential
I_S	Intensity of current source
N	Number count
φ	Potential function
ρ	Charge distribution
$f(x)$	Trial function
w	Line width
t	Line thickness
c	Wall to wall spacing
b_i	Height of the ith dielectric layer
ε_i	Permittivity of the ith dielectric layer
l	Length of the interconnect line
t	Thickness of the interconnect line
f	Frequency
d	Spacing between the interconnect line and the adjacent ground tracks
C_{Lower}	Capacitance per unit length of the region below the charge plane
C_{Upper}	Capacitance per unit length of the region above the charge plane
c'	Wall to wall spacing in the lower region
C_a	Line capacitance per unit length with dielectric replaced by air
L	Inductance per unit length of the interconnect line
v^a	Velocity of propagation

R	Resistance per unit length
ζ	Damping factor
t_d	50% delay time
t_r	90% rise time
t_o	Maximum/minimum overshoot time
t_s	Settling time
$\%O$	Maximum/minimum Percentage overshoot time
w_s	Width of the ground plane aperture
s	Edge to edge spacing between coupled lines
d_l	Spacing between interconnect line and ground tracks
pp'	Imaginary plane
C_{even}	Even-mode capacitance
C_{odd}	Odd-mode capacitance
G	Wall to wall spacing
Y_{even}	Even mode admittance
Y_{odd}	Odd mode admittance
Z_{even}	Even mode impedance
Z_{odd}	Odd mode impedance
C_v	Voltage coupling coefficient
k_c	Capacitive coupling coefficient
k_l	Inductive coupling coefficient
Z_{in}	Input impedance
$S_{11}, S_{12}, S_{21}, S_{22}$	S-parameters
V_p	Phase velocity

Chapter 1
Introduction

During the early evolutionary years of Integrated Circuit (IC) design, most of the research emphasized the need for performance improvement of transistors and other computing blocks. Transistor performance was seen as one on the most important design bottlenecks in those times. In 1965, Gordon E. Moore came up with a series of visionary predictions. Till date these predictions are valid to a large extent and the semiconductor industry is predominantly driven by these predictions. In a nutshell, Moore suggested that the performance and design complexity of ICs would grow exponentially and double every 18–24 months. During the same period, cost of transistors would reduce roughly by the same proportion each time we hit the Moore's next prediction [1]. Interestingly, while Gordon Moore thought it was only some observations, history of semiconductor growth only suggests that the notion of it being a law indeed holds good.

One of the important features of the Moore's prediction was the rate at which clock speeds has increased over the period of time. Starting from the modest MHz range in the 1960s, clock speeds have increased to more than a thousand times today. This means that the wires or interconnects connecting transistors on a chip, or chips on a board, or boards in a package have lengths that may be comparable to the wavelength of these signal frequencies. Also, it is worth mentioning that the scaling of feature sizes and its effect vis-à-vis the transmission of signals and clocks in high-speed interconnects have further aggravated signal transmission. Thus, the earlier notion of concentrating only on device improvement has no takers now. Increased transistor counts (from a few thousands to more than a billion), faster clocks (from a few MHz to several GHz) and skewed aspect ratios in every forthcoming technology node has scuttled overall system performance with signal integrity in high-speed interconnects being the foremost design challenge for the semiconductor industry [2–5].

Nowadays, electronic systems have tens of thousands of interconnect wires running from one end to the other in several layers stacked on top of each other. There is remarkable research focus on this topic; from developing newer

R. Sharma and T. Chakravarty, *Compact Models and Measurement Techniques for High-Speed Interconnects*, SpringerBriefs in Electrical and Computer Engineering, DOI: 10.1007/978-1-4614-1071-3_1, © The Author(s) 2012

Fig. 1.1 Simple illustration showing interconnect wire connecting on/off-chip devices

interconnect materials to inventing smarter process technologies to development of highly efficient Computer-Aided Design (CAD) algorithms. However, the most basic operation that an interconnect line does is to carry signal/clock from one point to another. A simple illustration of this fact is shown in Fig. 1.1. At the very core, we see interconnects as planar transmission lines that resemble microstrips or striplines. Later in the course of the text we shall explore the transmission line aspect in details. These transmission lines form the basic building blocks for developing compact models. Development of compact models for high-speed interconnects focuses on various aspects including extraction of RLCG parasitics, computing losses in dielectrics and metals, evaluation of crosstalk in coupled lines and equating signal transmission delay among other things. These analytical models are often verified by rigorous 2D or 3D field simulations available through commercial vendors and/or open literature available in the open source. Finally, time and frequency domain measurements are used to validate the analytical models.

1.1 Understanding High-Speed and High-Speed Effects

The term *High-Speed* is frequently used in interconnect design and modeling. It is related to the frequency of the signal as well as the physical dimensions, particularly length of interconnect. To understand this term from the interconnect perspective; it refers to the time taken by the signal to transmit from one end to the other end of the line [3, 6, 7]. The length of the interconnect (l) becomes particularly important, if the line length is of the order of one-tenth of the wavelength (λ) of the signal. Based on this comparison interconnects can be classified as electrically short or long. Subsequently the way they are modeled also varies. This is given in the following Table 1.1.

While signals do effectively transmit in an electrically shorter interconnect, the same does not hold true for longer wires. When electrical lengths exceed one-tenth of the signal wavelength we observe transmission line effects that makes it impossible for the line to be modeled as simple RC lumped model. Also, sharper rise/fall time ensures that the time to propagate the signal along the length of the wire is now comparable resulting in signal degradation. This is due to impedance mismatch; the impedance of interconnect simultaneously acts as the load to the

Table 1.1 Classification of interconnects based on their electrical lengths

Criteria	Type of interconnect	Model required
$l < \lambda/10$	Electrically short	Lumped
$l > \lambda/10$	Electrically long	Distributed

driver as well as the input impedance to the receiver, effectively isolating the driver from the receiver [3, 6]. These transmission line effects lead to signal overshoots, ringing and reflections that seriously degrade signal integrity.

According to authors in [5], interconnect hierarchy (or levels) is given below:

- Level 1: On-chip interconnects
- Level 2: Multichip module interconnects, packages
- Level 3: Printed circuit board interconnects
- Level 4: Backplane
- Level 5: Rack, connectors

Typically, the size of interconnects and the way they are modeled and analyzed vary from one level to another. Nowadays the performance of entire electronic systems is subject to the superior design and modeling of these interconnects. Signal integrity is a concern not only for high-speed data transfer but also for clock distribution networks in ICs and Network-on-Chip (NoCs) and System-on-Chip (SoCs) and vertical interconnects that are required in 3D integration [8–12].

Shrinking of technology has resulted in miniaturization of both devices and interconnects. Reduced aspect ratios mean smaller cross-sectional area, which in turn results in higher attenuation and makes data/clock recovery even tougher. There are two types of loss mechanisms in high-speed interconnects, namely skin effect and dielectric losses. Skin effect is a phenomenon where, at high frequencies, most of the current through a conductor is confined to a small cross-sectional area around the periphery of the conductor. As a result, the effective AC resistance per unit length of the conductor increases with frequency. Another interesting phenomenon has been the effect of dielectric losses on overall performance. Dielectric losses tend to be a function of frequency and are therefore more pronounced at higher frequencies [13–16]. Below is a summary of the various factors that determine high-speed effects in interconnects [3].

- Length of interconnect
- Cross-sectional dimensions
- Signal slew rates
- Clock speeds
- Skin effect losses
- Dielectric losses

High-speed effects or transmission line effects can cover a wide spectrum of phenomena ranging from propagation delay to crosstalk to reflections. Previously published literature provides exhaustive details on these effects and is a matter of continuous research in both academia and industry [3, 17–19]. Broadly, these effects are classified and listed below as:

- Delay
- Crosstalk
- Attenuation and losses
- Ringing and reflections

Digital signals can be summed up into two major constants, namely delay time (t_d) and rise time (t_r). The term t_d represents the finite amount of time required for the signal to transmit from one end of the interconnect line to another and is influenced by the parasitics of the line. Rise time (t_r) is particularly important when the t_r of the receiver is more than that of the source end, which leads to signal degradation. It should be noted by readers that these issues are extremely important from the perspective of digital design.

Closely spaced interconnect lines could result in one line becoming the driver while the other being the victim. The coupling between these two or more lines resulting due to mutual capacitances and inductances leads to crosstalk noise and possible alteration of logic levels in the victim line. As a design practice, designers employ intermediate shield lines to reduce crosstalk effects. However the problem may complicate owing to impedance mismatch of the signal lines due to these intermediate insertions [20]. Therefore one needs to address these issues in conjunction for the sake of achieving *optimum signal integrity*.

Attenuation in an interconnect system can be due to the losses in the metal (line or ground) or due to lossy dielectrics. Metal losses are predominantly skin effect losses at high frequencies where the conducting cross-sectional area of the line shrinks leading to increased line resistance and more signal attenuation. Also, it is worth mentioning the fact that losses in the ground plane cannot be simply ignored and are dictated by the return paths of the current flowing in the interconnect geometry. Dielectrics can be very lossy and therefore lead to losses that are remarkably higher at higher frequencies. One of the ways to reduce this loss is by using low-*k* materials. Recently, there has been a lot of research emphasis on the use of novel low-*k* materials including the use of air [21]. Low-*k* dielectrics also mean that we may have wider lines for retaining the same impedance values, thus resulting in even lower conductor losses. Air-clad interconnects could be seen as a revolutionary concept toward reducing these losses [21].

Signal reflections are primarily due to impedance mismatch in transmission lines. This can happen due to discontinuities in the transmission line that can be distributed or lumped in nature [3]. Common causes of discontinuities are transition between different levels of interconnects, orthogonal wiring, Through Silicon Vias (TSVs), solder bumps and wire bonding. In case of Printed Circuit Boards (PCBs), non-uniform transmission lines can be a major source of reflections and ringing. Signal degradation due to discontinuities and transitions is a major design headache in packages. Also, recent developments in 3D integration mean the use of more and more TSVs that results in a very complex and challenging situation for designers [22].

1.2 Evolution of Interconnects: Major Technological Developments

Compact interconnect models have evolved over the years primarily depending on the predictions of Moore's law. Signal and clock speeds and interconnect layout and geometry has governed the choice of analytical models. From modest lumped RC/RLC models to distributed frequency independent/dependent transmission line models to complex full-wave models that require solution of Maxwell's equation, modeling approach for high-speed interconnects has truly witnessed a sea change. Briefly, lumped RC/RLC models are valid for low frequencies (less than 1 GHz) where RC circuit models provide monotonic time response while ringing and overshoots can be incorporated using RLC models. However, as signal frequencies increase these models become inadequate to accurately predict the effects of signal degradation due to higher frequencies. Thus we require transmission line models that are based on quasi-TEM approximations that give distributed RLCG per unit length parameters [3, 6]. However, at higher frequencies (greater than 1 GHz), we see uneven current distribution that may result in skin effect, proximity effect and edge scattering [3, 6, 23] that requires frequency dependent parameter extraction models. Finally, beyond few GHz, 2D transmission line models are inaccurate due to Electromagnetic effects (EM) effects that can be correctly modeled using rigorous 3D full-wave Partial Element Equivalent Circuit (PEEC) models. The following part of this section illustrates the major technological changes in high-speed interconnects that have governed the choice of modeling approach for high-speed interconnects.

1.2.1 First Generation Interconnects: The RC Interconnect Era

Earlier research focused on simple lumped RC model that were quite accurate for the first generation interconnects. The primary purpose of these works was to compute delay in RC interconnects. One of the most preliminary works was presented by authors in [24] where RC tree networks were analyzed and propagation delay was computed for tree networks considering unit step inputs. Upper and lower bounds were estimated for general RC tree networks. This work was further extended by Lin and Mead in [25] where the delay value at any node in a general RC tree was computed using tree decomposition technique. For a two-port RC network various delay components were rigorously derived. More importantly the term delay was defined as Elmore's delay [26]; which essentially depends on the Laplace transform of the response. Thus for a monotonic output response $y(t)$, it is seen that the Elmore's delay (t_d), which is the centroid of $y'(t)$, closely matches the delay at each node of the tree structure. This is clearly explained in Fig. 1.2. Authors provide a very simple yet effective delay prediction model at any node in a general tree based on the path resistance and capacitances. The authors

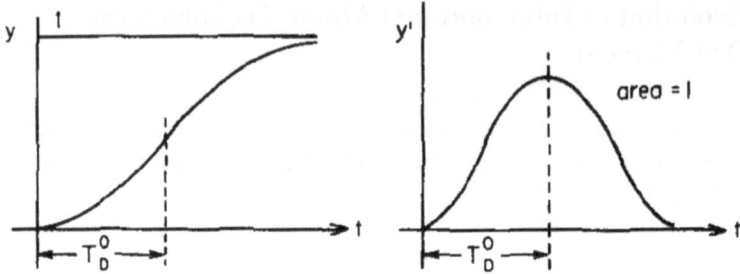

Fig. 1.2 Concept of Elmore's delay. © 1984 IEEE, reprinted, with permission, from Lin and Mead [25]

also established upper and lower delay bounds for these RC trees. These scholarly works laid a foundation for future works related to delay prediction in RC/RLC interconnects. However it is important to note that these models were more suitable for RC networks with monotonic response and had low accuracy. Further, inductive effects were not considered in these models due to the fact that Elmore's delay is essentially the first moment of the impulse response. Till late 1980s signal speeds were low (typically in MHz) ensuring that these models were widely used for lumped and distributed tree networks and interconnect lines. However, increased signal frequencies beyond a few GHz resulted in inductive effects and required modeling of RLC interconnects.

1.2.2 Second Generation Interconnects: Modeling Inductance

One of the earliest works highlighting the role of inductance and its effect on signal propagation was presented by Brews in [27]. RLC lines result in non-monotonic response and lead to signal overshoots and ringing besides introducing inductive delay component. This additional delay required modified Elmore's delay model for RLC interconnects as shown by authors [28]. For a typical RLC tree as shown in Fig. 1.3, signal characterization for a step input leads to the derivation of 50% delay, rise time, settling time and peak overshoots. The equivalent Elmore's delay model retains the same computational simplicity as its predecessor along with increased accuracy for computing delay at every node in the RLC tree for monotonic and non-monotonic responses. Higher signal speeds also resulted in crosstalk noise in coupled interconnect lines. A lot of literature is available on analyzing crosstalk related effects in closely spaced interconnects [29, 30]. Crosstalk in high-speed interconnects seriously degrade signal integrity resulting in logic failures in digital circuits. Designers therefore came up with modified layout designs and the use of shield lines to mitigate coupling between signal lines.

With the signal frequencies further increasing, lumped RLC models were revisited and transmission line models were proposed. The transmission line models, in turn, required EM and microwave effects and analysis of higher order modes.

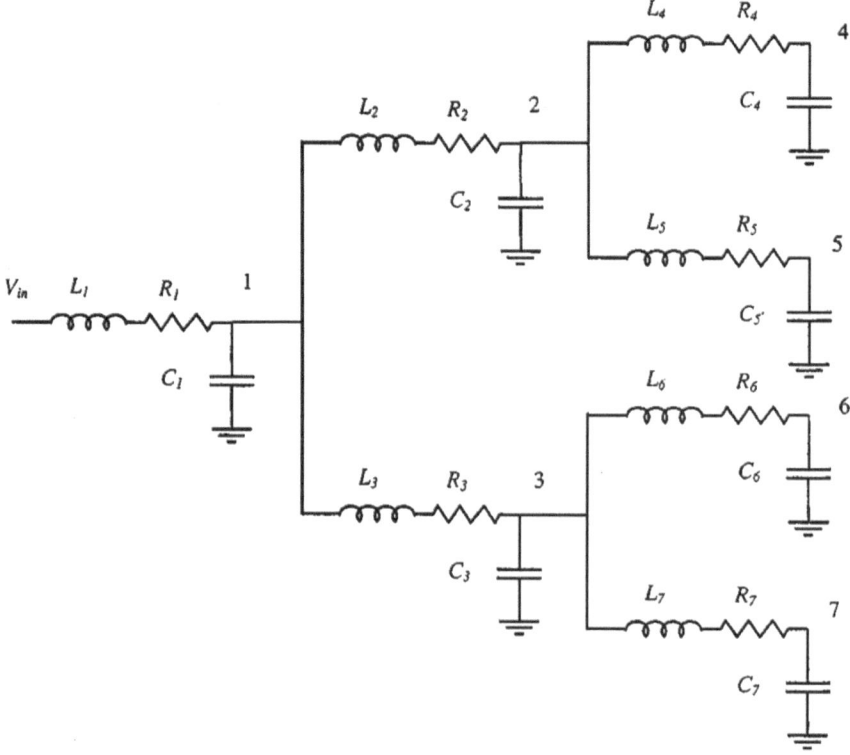

Fig. 1.3 RLC tree-based approximation. © 2000 IEEE, reprinted, with permission, from Ismail et al. [28]

This marked the coming of the next generation high-speed interconnect models with particular emphasis on exploring EM effects in these interconnects. Some of these models are elaborated by authors in a series of pioneering works in [31–34]. Present day interconnect models are complex and involve solution of transmission line properties for evaluation of crosstalk, delay, ringing and reflection. With the International Technology Roadmap for Semiconductors (ITRS) predictions on signal frequencies well beyond a few GHz, it is envisaged that the development of compact models will require even deeper understanding of circuit theory, EM theory and microwave engineering. Thus we are witnessing a merger of several fundamental fields of electrical engineering with signal integrity coming to the forefront.

1.2.3 Copper and Low-k Interconnects

Since the beginning of integrated circuit technology Al–SiO$_2$ as conductor-substrate pair has been a natural material choice. The transition to copper (Cu) as a reliable interconnect material has by far been the greatest paradigm shift in

Table 1.2 Typical resistivity of metals

Metal	Bulk resistivity [$\mu\Omega$-cm]
Silver	1.63
Copper	1.67
Gold	2.35
Aluminum	2.67
Tungsten	5.65

Fig. 1.4 Copper versus aluminum as an interconnect material. © 1995 IEEE reprinted, with permission, from Bohr [36]

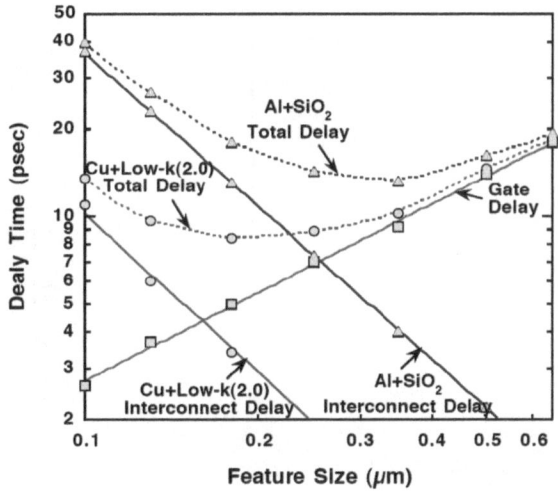

interconnect technology. The fundamental limitation of Al–SiO$_2$-based interconnects became evident due to higher resistance and capacitance offered by Al lines and SiO$_2$ substrates that resulted in higher RC wiring delay. Thus great impetus was laid on the development of Cu and low-k-based interconnects. The resistivity of Cu is typically 40% less than that of Al as shown in Table 1.2. This results in lesser RC wiring delay as shown in Fig. 1.4. Since the signal delay is proportional to the resistivity of the interconnect material, we can see 40% reduction in the RC time constant when using Cu over Al. When combined with low-k material the RC delay can decrease to 50% as compared to that of Al–SiO$_2$. Also, copper exhibits better electromigration properties and is therefore more reliable as an interconnect material. This is due the fact that Cu has 10–100 times more resistance to electromigration effects than Aluminum [35]. Use of Cu can minimize process steps and reduce interconnect routing and interconnect levels thus resulting in cost savings and better yield.

However, the most significant challenge in using Cu interconnects lie in the processing of this material. As against depositing Al over the entire wafer and then patterning the lines using reactive ion etching, the use of Cu interconnects requires an altogether new processing approach to pattern Cu. The Damascene process [37] has emerged as a reliable process standard to pattern Cu interconnects. Readers are advised to review the extensive literature available on the processing of Cu interconnects for a better understating on this topic. However, from the modeling perspective, Cu interconnects posed several interesting challenges; in that modeling of line resistance and inductance had to be revisited. The deposition of Cu over silicon results in Cu diffusing inside the silicon substrate, which leads to "deep-level defects" [38]. To overcome this we need a thin barrier layer followed with a passivation layer, if necessary, that isolates the Cu from silicon. Importantly, the scaling of barrier layer is not proportional to that of the metal layer. Thus the effective resistivity of the Cu interconnects remains higher than its bulk resistivity because of the higher resistance offered by barrier layer. Two important effects are worth noting here: the dishing effect that refers to loss of Cu in wider lines and erosion effect that results in loss of dielectric material between metal lines. These two effects lead to higher resistance and capacitance. Also, with continuous scaling of Cu interconnects electron scattering at the surface and Cu grain size need to be considered for efficient modeling of these interconnects.

Since Cu has lower resistivity the effect of line inductance is even more significant. At higher frequencies this inductive effect dominates causing signal overshoots and ringing and reflections due to impedance mismatch. This requires modeling of self and mutual inductance terms that were previously unknown to designers using Al–SiO$_2$ lines. The interconnect lines can now be viewed as second-order system requiring recalculation of equivalent Elmore's delay [27, 28] and deeper understanding of return paths. The calculation of return path current is particularly complex and may require consideration of neighboring and/or distant signal/ground tracks depending on the frequency of interest.

While there could be significant reduction in resistance using Cu interconnects similar reduction in capacitance can be achieved using low-k dielectrics. Thus with a Cu-low-k combination we could achieve lower RC delay for the same wire aspect ratio. This forms the basic motivation for the use of various low-k materials. Typically any low-k material should have all or some these characteristics [39]:

- Electrical: relative permittivity (ε_r) <3; isotropic
- Mechanical: excellent adhesion with metals, dielectrics etc.
- Thermal: low thermal expansion and high thermal stability
- Commercial: environmentally safe and cost effective

There are several types of low-k dielectrics; namely inorganic (glass), organic (polyimide, Teflon), hybrid (polymers) and porous [39]. However, in the recent years there has been an attempt to use ultra-low-loss air dielectrics. While it is not possible to have complete air-clad interconnects for apparent mechanical reasons, a hybrid design is proposed by authors in [21, 40]. The primary reason for the use of air-clad interconnects is to have minimum energy dissipation at frequencies well into a few of tens of GHz. At these frequencies dielectric losses dominate over conductor losses and air-clad interconnects can provide extremely low-loss planar and vertical links for chip–chip interconnections. From modeling point of view, development of models for calculating losses and thereby predicting energy-delay product is most important. Air-clad interconnects present the extreme limits to the use of low-k dielectrics and loss models can be used to benchmark future optical interconnects [21]. Readers should expect interesting research in this area in the days to come.

1.2.4 Optical Interconnects

Electrical interconnects including Cu-low-k interconnects have fundamental limitations in terms of signal attenuation, bandwidth and crosstalk. The attenuation increases with frequency as explained in the previous section. Combined with inter-symbol interference and crosstalk from neighboring wires, impedance mismatches and package losses signal integrity can be seriously compromised. Even if we use equalization and low-k materials to alleviate losses bandwidth in these high-speed links could still be a bottleneck. Optical interconnects can come to our rescue in such a scenario. It can provide extremely high bandwidths with much lower power and area requirements [41–43]. Some of the key benefits of optical interconnects over electrical interconnects include:

- Absence of electromagnetic effects (impedance mismatch, crosstalk and inductive effects)
- Distance independence of optical interconnects
- Frequency independence of optical interconnects

However, one needs a greater understanding of photonics, transmitter and waveguide delay and receiver power and delay modeling to effectively model optical interconnects. The field of optical interconnects has shown remarkable progress over the years and is a subject matter of continuous research and development. Some of the technical challenges in this area include design issues with receiver circuits, optoelectronic devices, integration issues and cost effectiveness. Design and modeling of chip–chip optical interconnect systems is however beyond the scope of this book. Readers are referred to literature available in [41–47] for both historical as well as current perspective on this research activity.

1.3 Concluding Remarks

This chapter briefly introduces the fundamentals of interconnects and issues related to compact models for interconnects to our readers. There is vast amount of literature available on the above topics and readers are encouraged to refer to them. The performance of interconnects present a fundamental bottleneck in overall system performance. While there are several accurate commercial simulators available, development of compact models is imperative to design space exploration and obtaining physical insights. In a nutshell, this chapter lays the basic building block for more detailed analysis to be followed in the subsequent chapters.

In the next chapter, we present comparative summary of some of the analytical approaches that can be used to develop compact physical models for high-speed interconnects. In particular we emphasize on the unified approach and present its application in developing compact models for high-speed interconnects. Later we present several illustrations to highlight the versatility of this approach for modeling high-speed interconnects leaving a window of opportunity for its application in more complex interconnect networks.

References

1. G.E. Moore, Cramming more components onto integrated circuits. Electronics **38**(8), 114–117 (1965)
2. M. Horowitz, K.Y. Chih-Kong, S. Sidiropoulos, High-speed electrical signaling: overview and limitations. IEEE Micro **18**(1), 12–24 (1998)
3. R. Achar, M.S. Nakhla, Simulation of high-speed interconnects. Proc. IEEE **89**(5), 693–728 (2001)
4. S.H. Hall, G.W. Hall, J.A. McCall, *High-Speed Digital System Design—A Handbook of Interconnect Theory and Design Practices* (IEEE-Wiley Press, New York, 2000)
5. T.C. Edwards, M.B. Steer, *Foundations of Interconnect and Microstrip Design* (Wiley, New York, 2000)
6. C.R. Paul, *Analysis of Multiconductor Transmission Lines* (Wiley, New York, 1994)
7. H.B. Bakoglu, *Circuits, Interconnections and Packaging* (Addison Wesley Publishers, Reading, 1990)
8. E.G. Friedman, Clock distribution networks in synchronous digital integrated circuits. Proc. IEEE **89**(5), 665–692 (2001)
9. R. Ho, K.W. Mai, M.A. Horowitz, The future of wires. Proc. IEEE **89**(4), 490–504 (2001)
10. W.R. Davis et al., Demystifying 3D ICs: the pros and cons of going vertical. IEEE Des. Test Comput. **22**(6), 498–510 (2005)
11. J.A. Davis et al., Interconnect limits on gigascale integration (GSI) in the 21st century. Proc. IEEE **89**(3), 305–324 (2001)
12. A. Roy, M.H. Chowdhury, RF/Wireless interconnects in future on-chip and board-level clock distribution network, in *Proceedings of the IEEE International Conference on Electro/Information Technology* (2007), pp. 542–545
13. K. Banerjee, A. Mehrotra, Global (interconnect) warming. IEEE Circuits Devices Mag. **17**(5), 16–32 (2001)
14. H. Mavoori, Copper/low-*k* interconnects for smaller and faster circuits. J. Miner. Met. Mater. Soc. **51**, 36 (1999)
15. H. Cho, P. Kapur, K.C. Saraswat, Power comparison between high-speed electrical and optical interconnects for interchip communication. J. Lightwave Technol. **22**(9), 2021–2033 (2004)
16. W.J. Dally, J.W. Poulton, *Digital Systems Engineering* (Cambridge University Press, Cambridge, 1998)
17. B. Young, *Digital Signal Integrity: Modeling and Simulation with Interconnects and Packages* (Prentice Hall, New Jersey, 2001)
18. M.S. Nakhla, Q.J. Zhang, *Modeling and Simulation of High Speed VLSI Interconnects* (Springer, Netherlands, 1994)
19. W.-K. Chen, *The VLSI Handbook* (CRC Press, New York, 2000)

20. R. Sharma, T. Chakravarty, A.B. Bhattacharyya, Analytical model for optimum signal integrity in PCB interconnects using ground tracks. IEEE Trans. Electromagn. Compat. **51**(1), 67–77 (2009)
21. V. Kumar, R. Bashirullah, A. Naeemi, Modeling, optimization and benchmarking of chip-to-chip electrical interconnects with low loss air-clad dielectrics, in *Proceedings of the Electronic Components and Technology Conference* (2011), pp. 2084–2090
22. C. Liu, S.K. Lim, A study of signal integrity issues in through-silicon-via-based 3D ICs, in *Proceedings of the International Interconnect Technology Conference* (2010), pp. 1–3
23. A.R. Djordjević, T.K. Sarkar, Closed-form formulas for frequency-dependent resistance and inductance per unit length of microstrip and strip transmission lines. IEEE Trans. Microw. Theory Tech. **42**(2), 241–248 (1994)
24. J. Rubinstein, P. Penfield, M.A. Horowitz, Signal delay in RC tree networks. IEEE Trans. Comput. Aided Des. Integr. Circuits Syst. **2**(3), 202–211 (1983)
25. T. Lin, C. Mead, Signal delay in general RC networks. IEEE Trans. Comput. Aided Des. Integr. Circuits Syst. **3**(4), 331–349 (1984)
26. W.C. Elmore, The transient response of damped linear networks with particular regard to wideband amplifiers. J. Appl. Phys. **19**(1), 55–63 (1948)
27. J.R. Brews, Overshoot controlled RLC interconnections. IEEE Trans. Electron Devices **38**(1), 76–87 (1991)
28. Y.I. Ismail, E.G. Friedman, J.L. Neves, Equivalent Elmore's delay for RLC trees. IEEE Trans. Comput. Aided Des. Integr. Circuits Syst. **19**(1), 83–97 (2000)
29. T. Sakurai, Closed-form expressions for interconnection delay, coupling, and crosstalk in VLSI's. IEEE Trans. Electron Devices **40**(1), 118–124 (1993)
30. A. Vittal, M. Marek-Sadowska, Crosstalk reduction for VLSI. IEEE Trans. Comput. Aided Design Integr. Circuits Syst. **16**(3), 290–298 (1997)
31. J.A. Davis, J.D. Meindl, Compact distributed RLC interconnect models—part I: single line transient, time delay, and overshoot expressions. IEEE Trans. Electron Devices **47**(11), 2068–2077 (2000)
32. J.A. Davis, J.D. Meindl, Compact distributed RLC interconnect models—part II: coupled line transient expressions and peak crosstalk in multilevel networks. IEEE Trans. Electron Devices **47**(11), 2078–2087 (2000)
33. R. Venkatesan, J.A. Davis, J.D. Meindl, Compact distributed RLC interconnect models—part III: transients in single and coupled lines with capacitive load termination. IEEE Trans. Electron Devices **50**(4), 1081–1093 (2003)
34. R. Venkatesan, J.A. Davis, J.D. Meindl, Compact distributed RLC interconnect models—part IV: unified models for time delay, crosstalk and repeater insertion. IEEE Trans. Electron Devices **50**(4), 1094–1102 (2003)
35. P. Kapur, J.P. McVittie, K.C. Saraswat, Technology and reliability constrained future copper interconnects—part I: resistance modeling. IEEE Trans. Electron Devices **49**(4), 590–597 (2002)
36. M.T. Bohr, Interconnect scaling—the real limiter to high performance ULSI, in *Proceedings of the 1995 IEEE International Electron Devices Meeting* (1995), pp. 241–244
37. C. Andricacos, C. Uzoh, J.O. Dukovic, J. Horkans, H. Deligianni, Damascene copper electroplating for chip interconnections. IBM J. Res. Dev. **42**(5), 567 (1998)
38. N.D. Arora, Modeling and characterization of copper interconnects for SoC design, in *Proceedings of the Simulation of Semiconductor Processes and Devices* (2003), pp. 1–6
39. M. Clarke, Introducing low-*k* dielectrics into semiconductor processing. *Application Note: MAL123*, Mykrolis (2003)
40. T.J. Spencer, P.J. Joseph, T.H. Kim, M. Swaminathan, P.A. Kohl, Air-gap transmission lines on organic substrates for low-loss multiprocessor interconnects. IEEE Trans. Microw. Theory Tech. **55**(9), 1919–1925 (2007)
41. H. Cho, P. Kapur, K.C. Saraswat, Power comparison between high-speed electrical and optical interconnects for interchip communication. J. Lightwave Technol. **22**(9), 2021–2033 (2004)

42. P. Kapur, K.C. Saraswat, Optical interconnects for future high performance integrated circuits. Phys. E Low Dimens. Syst. Nanostruct. **16**(3–4), 620–627 (2003)
43. D.A.B. Miller, Rationale and challenges for optical interconnects to electronic chips. Proc. IEEE **88**(6), 728–749 (2000)
44. R.H. Havemann, J.A. Hutchby, High-performance interconnects: an integration overview. Proc. IEEE **89**(5), 586–601 (2001)
45. S. Luryi, J. Xu, A. Zaslavsky, Ultrafast nanophotonic devices for optical interconnects, in *Future Trends in Microelectronics: From Nanophotonics to Sensors to Energy* (2010), pp. 43–48
46. www.zurich.ibm.com/st/photonics/interconnects
47. *The 50G Silicon Photonics Link,* Intel White Paper. www.intel.com

Chapter 2
Compact Modeling of High-Speed Interconnects

High-speed interconnects are essentially planar transmission lines. The fundamental mode of propagation in transmission line interconnects is the transverse electromagnetic (TEM) wave [1]. In ideal case, when the conductivity of the line is infinity the basic mode of propagation would be the TEM mode. This is assuming that the medium in which the line is embedded is considered to be homogeneous, lossless and isotropic. However for most practical cases, the lines have finite conductivity that results in a deviation from the TEM mode. The properties of the dielectric material are also far from ideal with dielectric losses dominating conductor losses as frequency scales up. Therefore, interconnect lines embedded in inhomogeneous substrates cannot support pure TEM mode. The modified mode of propagation has small axial components of the electric and magnetic fields. The field distribution in such a non-ideal transmission line interconnect closely represents the ideal TEM mode with negligible electric/magnetic field components and is called the quasi-TEM mode [1, 2]. Transmission line theory has two aspects: In one case, the propagation of electromagnetic waves is studied when the characteristic parameters of the line are given. In the other case, the conductor geometry is known and the line parameters such as the characteristic impedance, attenuation constant, propagation constant and the shunt capacitance are to be obtained. This aspect is particularly suited for interconnect design and analysis. With the quasi-TEM approximation, the calculation of these line parameters requires the solution of the two-dimensional Laplace's equation. This solution is based on the computation of the boundary conditions governed by the geometry of the line.

There are many analytical techniques available in the literature for the solution of the Laplace's equation. This chapter presents a qualitative overview of some of the most commonly used analytical techniques, which includes, among others, the conformal transformation method [1, 3–6], the finite-difference method [7–10] and the variational method [1, 2, 11–13]. In the following section, a comparative overview of these techniques is presented. In the latter half of this chapter, we propose the application of a unified approach for analysis of high-speed

R. Sharma and T. Chakravarty, *Compact Models and Measurement Techniques for High-Speed Interconnects*, SpringerBriefs in Electrical and Computer Engineering, DOI: 10.1007/978-1-4614-1071-3_2, © The Author(s) 2012

interconnects. Based on the comparative summary of the major analytical approaches and the relative advantages of this unified approach; we feel that this technique is well suited for planar chip–chip interconnects. Derivation of line parameters using a unified approach that combines the variational method with the transverse transmission line technique [2, 14] is presented. This chapter makes an attempt to highlight the relative advantages and applicability of the unified approach and concludes that this technique fits in our analytical models better than other available methods. Although the comparison presented in the next section is qualitative only, it would provide reasonable insight to the reader; further leaving a scope for the employment of above-listed techniques to be used in the modeling of high-speed interconnects.

2.1 Review of the Analytical Methods

A lot of research has been presented in the past on transmission line modeling. Most of the earlier works focused on problems in *MIC*s and other microwave circuits. However, with the signal frequencies now entering well into the GHz range, it is only appropriate to apply some of these analytical techniques to address signal integrity issues in high-speed VLSI and chip–chip interconnects. The techniques available to solve TEM and quasi-TEM problems can be broadly classified into two classes, namely exact and approximate [15]. Some of the most common approaches under these two classes are shown in Fig. 2.1.

Since most of these techniques are extensively covered in the published research on microwave theory; we will only highlight their relative merits and drawbacks with an objective to justify the choice of the method of analysis used in this book. The formulae obtained using conformal mapping are derived using the Schwartz-Christoffel conformal transformation. This method enables one to evaluate the capacitances and characteristic impedance between straight-sided conductors when the problem can be reduced to two dimensions, for example the cross-section plane of a transmission line. The boundary of the cross-section is transformed into a simpler boundary for which the solution is known using transformations in the complex plane. Finally the capacitance and characteristic impedance of the original boundary are equal to the respective quantities of the transformed boundary [4, 5]. The conformal transformation technique is exact and accurate. Since, the capacitance term of each element retains the correct dependencies on the line geometry; we require very few 2D simulations resulting in lesser computational time. However, in case of transmission lines with inhomogeneous medium the application of conformal transformation may become quite complicated. To overcome this problem, the Discrete Variational Conformal (*DVC*) transformation method suggested by Diaz [15] seems to be more suitable in cases where the geometry of the structure under analysis is not simple. However, in case of microstrip-like interconnects (that are commonly encountered in chip–chip interconnects) the DVC method provides results that are virtually identical.

Analytical approaches

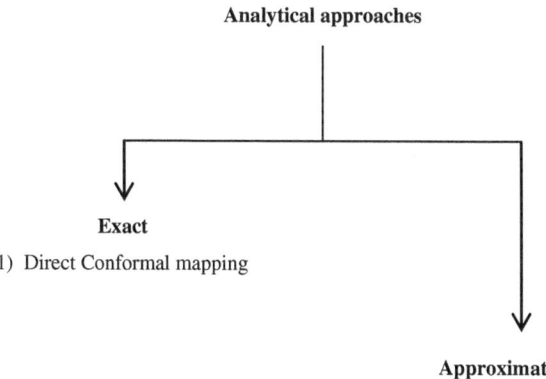

Exact

1) Direct Conformal mapping

Approximate

1) Conformal mapping [5];
2) Discrete variational conformal technique [15]
3) Finite-difference method [9,10]
4) Method of moments [16]
5) Variational approach [2, 11, 14]
6) Spectral domain or full-wave method [17,18]
7) Finite element method
8) Variational series method [20]

Fig. 2.1 Classification of analytical methods. © 1986 IEEE reprinted, with permission, from Diaz [15]

   Finite-difference method is also widely used for analysis of planar transmission lines. The application of the finite-difference method to TEM transmission lines involves the solution of Laplace's equation in the cross-sectional plane subject to boundary conditions on the inner and the outer conductors. The entire domain between the conductors is divided into a finite set of mesh points. Laplace's equation is then solved in the finite-difference form by digital computation [14]. This method can be applied to TEM lines and has been extended to quasi-TEM transmission lines with limited inhomogeneity and is elaborately explained in [7, 8]. However, the finite-difference technique is vastly limited to homogeneous and geometrically simpler structures. With complex interconnect layout designers do not have the luxury to assume such simplifications. The accuracy of the method depends on the fineness of the mesh size (as in coupled strip transmission lines). This results in very large set of equations to be solved, leading to the problem of convergence and therefore inaccuracy. Kammler's method given in [16] can be used to analyze interconnects with multiple layers of dielectrics, but it may prove to be computationally cumbersome. Other techniques like the finite element method and the spectral domain analysis also suffer from problems in analyzing open microstrip cases and thus have limited applications. In case of open multi-dielectric planar lines, the application of finite elements leads to two difficulties; namely.

- the infinite field extension due to the open structure,
- and, the field singularities caused by the conductor edges.

A combined approach making use of the variational series based on the conformal transformation has been reported by Smith [19]. This method overcomes the difficulties of convergence and singularities encountered in the finite-difference method and/or finite elements.

Finally, the variational method [2, 11, 14] is generally applied to those problems where the physical system under study acts so that some function of its behavior attains the least or the greatest value. The variational method can be used to obtain the expression for line capacitance of a transmission line in an inhomogeneous, isotropic/anisotropic media. This geometrical environment is exactly the case in high-speed interconnects. When combined with the transverse transmission line technique of determining the Green's function [7, 8, 20], line parameters can be computed for a variety of structures. The method is simple and generalized due to the ease of computing Green's function using the transverse transmission line technique and gives fairly accurate results without much computational effort. This method has certain limitations also; namely

- dielectric material should be of low loss,
- the method assumes a TEM mode and neglects radiation effects,
- and, the accuracy of the results depends on the trial function.

In case of the modern day interconnect design the above-mentioned points are largely taken care of. Also, the trial function can be chosen after experimental verification leaving lesser scope for inaccuracy. To summarize the above discussion, the authors feel that the variational analysis in the space domain combined with the transverse transmission line technique offers a robust approach for analysis of high-speed transmission line interconnects. The conformal mapping technique can be very complicated in the case of inhomogeneous interconnects. Modern IC layouts cannot certainly guarantee homogeneity. Also, the finite-difference method involves a numerical evaluation and is thus limited to simpler structures. The other listed techniques have even less applicability than these methods. The variational method—though approximate—offers a simpler way of determining propagation parameters including line parasitics. When combined with the transverse transmission line technique of determining the Green's function [7, 8, 15], the derivation for the capacitance of the interconnect line becomes quite simple and reasonably accurate and is therefore suited for CAD applications. Since the variational method treats the dielectric boundary conditions in a generalized way, it is possible to analyze multilayer interconnect lines. The accuracy of this method is insensitive to the choice of the trial function (discussed in the following sections). Authors in [14] suggest that it is possible to take into account all the dielectric boundary conditions no matter how many planar boundaries exist in these lines. The method is based on the calculation of the line capacitance by the static field theory and therefore is an approximation to EM theory. It is felt that unlike conformal mapping and other mentioned techniques—which are also static field theories—the analytical treatment of multiple boundaries is easier by the variational method [2, 14]. The computational time is also far less than for other techniques which makes it suitable for CAD-related applications.

The above discussion illustrates the possible application of variational method combined with the transverse transmission line technique for the analysis of the chip–chip interconnects. However, interested readers are strongly encouraged to read the literature presented by authors in [7, 14, 20]. It is felt that the method is explained in detail in these literatures.

2.2 Unified Approach

Classically speaking the unified approach refers to the variational analysis combined with the transverse transmission line technique. The approach was first reported for analysis of Microwave Integrated Circuit (MIC), Monolithic Microwave Integrated Circuit (MMIC) and planar transmission lines. In this approach, the expression for the capacitance of a transmission line is determined by the variational technique. The Green's function is computed using the transverse transmission line technique in the space domain. In this section, derivation for the line capacitance using unified approach is reproduced from the historical literatures for both single as well as coupled line structures.

2.2.1 Computing Green's Function

Let us assume a unit charge located at (x_0, y_0) as shown in Fig. 2.2. The Green's function should satisfy the Poisson's differential equation in the $x-y$ plane and is given by:

$$\nabla_t^2 G(x, y/x_o, y_o) = -\frac{1}{\varepsilon} \delta(x - x_o) \cdot \delta(y - y_o) \tag{2.1}$$

For an interconnect line over a multilayered substrate, the boundaries at the interface of the dielectrics are given by:

$$G(x, s_{j-0}) = G(x, s_{j+0}) \tag{2.2}$$

$$\varepsilon_j \frac{\partial G(x, s_{j-0})}{\partial y} = \varepsilon_{j+1} \frac{\partial G(x, s_{j+0})}{\partial y} \tag{2.3}$$

Figure 2.3a, b represent the configuration corresponding to a microstrip line with rectangular side walls and a corresponding geometry required to compute the Green's function.

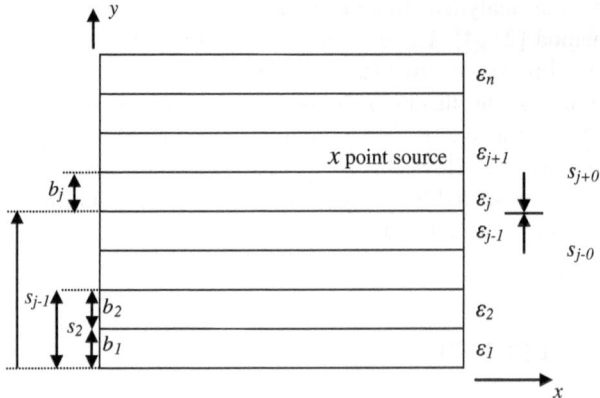

Fig. 2.2 Geometry of an n-layer dielectric with side walls and a point charge at (x_0, y_0). © 1978 IEEE reprinted, with permission, from Crampagne et al. [20]

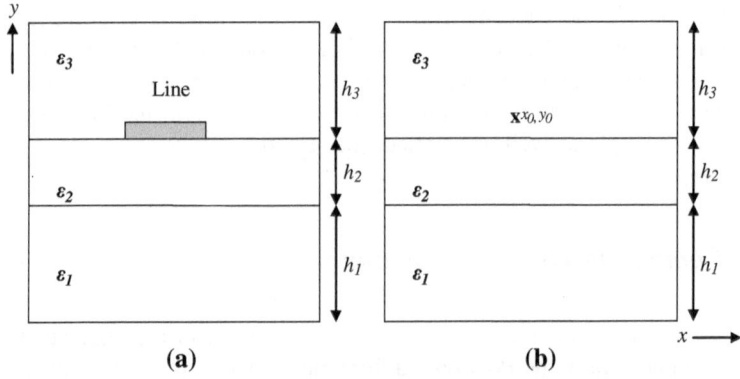

Fig. 2.3 a Microstrip line with rectangular side walls. **b** Geometry to calculate the Green's function. © 1978 IEEE reprinted, with permission, from Crampagne et al. [20]

Figure 2.3 represents only a particular case and the number of cases depends on the boundary conditions at the rectangular walls. The boundary conditions satisfied at the vertical walls can be either of the Dirichlet type (electric wall, $G = 0$) or of the Neumann type (magnetic wall, $\partial G/\partial n = 0$). The boundary conditions on the lower and upper surfaces (or horizontal walls) can be taken into account using the transverse transmission line technique and will be discussed later. If we assume any arbitrary conditions on the horizontal walls, then there could be three specific cases of boundary conditions on the vertical walls as shown in Fig. 2.4.

The Green's function can be expressed as the summation of the product of elementary functions given below:

$$G = \sum_n G_n(x)G_n(y) \tag{2.4}$$

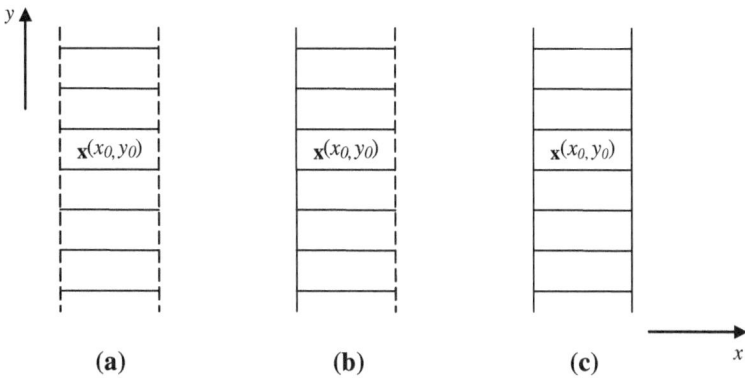

Fig. 2.4 Magnetic (*solid line*) and electric (*dashed line*) boundaries. © 1978 IEEE reprinted, with permission, from Crampagne et al. [20]

We should now derive the identities $G_n(x)$ and $G_n(y)$. In order to satisfy the boundary conditions on the vertical walls separated by wall spacing c, the following expressions are found for $G_n(x)$ for three separate cases corresponding to Fig. 2.4.

Case a Electric wall at $x = 0$ and c:

$$G_n(x) = \sin\frac{n\pi x}{c}, \qquad n = 1, 2, \ldots, \infty \tag{2.5}$$

Case b Electric wall at $x = 0$ and magnetic wall at $x = c$:

$$G_n(x) = \sin\frac{(2n+1)\pi x}{2c}, \qquad n = 0, 1, 2, \ldots, \infty \tag{2.6}$$

Case c Magnetic walls at $x = 0$ and c:

$$G_n(x) = \cos\frac{n\pi x}{c}, \qquad n = 2, 3, \ldots, \infty \tag{2.7}$$

It is seen that the functions, $\sin(n\pi x/c)$, $\sin[(2n+1)\pi x/2c]$ and $\cos(n\pi x/c)$ are orthogonal in the interval $(0, c)$. Substituting the expressions in (2.5)–(2.7) the following differential equations are obtained:

Case a

$$\left[\frac{d^2}{dy^2} - \left(\frac{n\pi}{c}\right)^2\right]G_n(y) = -\frac{2}{c\varepsilon}\delta(y - y_o)\sin\frac{n\pi x_o}{c} \tag{2.8}$$

Case b

$$\left[\frac{d^2}{dy^2} - \left(\frac{(2n+1)\pi}{2c}\right)^2\right]G_n(y) = -\frac{2}{c\varepsilon}\delta(y - y_o)\sin\frac{(2n+1)\pi x_o}{2c} \tag{2.9}$$

Case c

$$\left[\frac{d^2}{dy^2} - \left(\frac{n\pi}{c}\right)^2\right] G_n(y) = -\frac{2}{c\varepsilon} \delta(y - y_o) \cos\frac{n\pi x_o}{c} \qquad (2.10)$$

The Green's function, G and $G_n(y)$, should satisfy the boundary conditions at the various dielectric interfaces given by Eqs. (2.2) and (2.3).

2.2.2 Transverse Transmission Line Technique

Having derived $G_n(x)$ in the last section we now compute the identity $G_n(y)$. The Transverse transmission line technique provides a simpler method that numerically evaluates the Green's function. For N number of dielectric layers, the solution of above differential equations leads to a set on linear equations with $2N$ number of rows. Consider a transmission line with a current source of intensity I_s at the charge plane $y = y_0$. The voltage and current relations along the line are found to be:

$$\frac{dV}{dy} = -\gamma Z I_s \qquad (2.11)$$

$$\frac{dI}{dy} = -\frac{\gamma}{Z} V + I_s \delta(y - y_o) \qquad (2.12)$$

Here, Z is the characteristic impedance of the line and γ is the propagation constant. Solving (2.11) and (2.12), we get:

$$\frac{d^2 V}{dy^2} - \gamma^2 V = -\gamma Z_c I_s \delta(y - y_o) \qquad (2.13)$$

If we consider the transmission line as a set of stepped characteristic admittances in parallel with Y_{cj} as the characteristic admittance of the jth section of transmission line, then the continuity conditions at the interfaces between the two differential admittances are given by:

$$V_j = V_{j+1} \qquad (2.14)$$

and $I_{-j} = I_{j+1}$. Thus we get:

$$Y_{cj} \frac{\partial V_j}{\partial y} = Y_{cj+1} \frac{\partial V_{j+1}}{\partial y} \qquad (2.15)$$

Comparing Eqs. (2.13), (2.14) and (2.15), author in [14] came up with the following similarities:

1. The functions characterizing the Green's function can be identified by the voltage along the line.

$$V \equiv G_n(y) \tag{2.16}$$

2. The dielectric constant of the jth layer can be identified by the characteristic admittance of the transmission line for that section.

$$Y_{cj} = \varepsilon_j \tag{2.17}$$

Thus, the boundary conditions satisfied by the Green's function at the various dielectric interfaces are equivalent to the boundary conditions satisfied by the voltages at the interfaces between two dissimilar characteristic admittances. The voltage on the transmission line at $y = y_o$ is given by:

$$V|_{y=y_o} = \frac{I_s}{Y} \tag{2.18}$$

where Y is the admittance at $y = y_o$. We can now obtain the Green's function for the three cases listed in Fig. 2.4 as given by [14]:

Case a

$$Z = \frac{1}{\varepsilon}, \quad \gamma = \frac{n\pi}{c}, \quad \text{and} \quad I_s = \frac{2}{n\pi} \sin \frac{n\pi x_o}{c}$$

$$G_n(y)|_{y=y_o} = \frac{2}{n\pi Y} \sin \frac{n\pi x_o}{c} \tag{2.19}$$

Substituting (2.5) and (2.19) in (2.4), the Green's function at the charge plane $y = y_o$ becomes:

$$G(x, y/x_o, y_o)|_{y=y_o} = \sum_{n=1}^{\infty} \frac{2}{n\pi Y} \sin \frac{n\pi x}{c} \sin \frac{n\pi x_o}{c} \tag{2.20}$$

Case b

$$Z = \frac{1}{\varepsilon}, \quad \gamma = \frac{(2n+1)\pi}{2c}, \quad \text{and} \quad I_s = \frac{4}{(2n+1)\pi} \sin \frac{(2n+1)\pi x_o}{2c}$$

$$G_n(y)|_{y=y_o} = \frac{4}{(2n+1)\pi Y} \sin \frac{(2n+1)\pi x_o}{2c} \tag{2.21}$$

Substituting (2.6) and (2.21) in (2.4), the Green's function at the charge plane $y = y_o$ becomes:

$$G(x, y/x_o, y_o)\big|_{y=y_o} = \sum_{n=0}^{\infty} \frac{4}{(2n+1)\pi Y} \sin \frac{(2n+1)\pi x}{2c} \sin \frac{(2n+1)\pi x_o}{2c} \qquad (2.22)$$

Case c

$$Z = \frac{1}{\varepsilon}, \quad \gamma = \frac{n\pi}{c}, \quad \text{and} \quad I_s = \frac{2}{n\pi} \cos \frac{n\pi x_o}{c}$$

$$G_n(y)\big|_{y=y_o} = \frac{2}{n\pi Y} \cos \frac{n\pi x_o}{c} \qquad (2.23)$$

Substituting (2.7) and (2.23) in (2.4), the Green's function at the charge plane $y = y_o$ becomes:

$$G(x, y/x_o, y_o)\big|_{y=y_o} = \sum_{n=1}^{\infty} \frac{2}{n\pi Y} \cos \frac{n\pi x}{c} \cos \frac{n\pi x_o}{c} \qquad (2.24)$$

We can now see that the Green's function can be deduced from the admittance Y that can be obtained using the standard transmission line admittance equation. As mentioned earlier the unified approach is a combination of the variational technique and the transverse transmission line technique. In that the Green's function is computed using the latter as shown in the preceding discussion. Table 2.1 gives the identification of all the characteristic parameters concerned in the above discussion.

2.2.3 Variational Method

In the unified approach the variational method is used to compute the capacitance per unit length [1]. Let us consider a system of perfect conductors S_1, S_2, ... S_N with Q_1, Q_2,... Q_N as the charges on the conductors and V_1, V_2,..., V_N as the potential difference. The potential function φ in the space domain happens to be the solution of the Laplace's equation. The electrostatic energy stored in such a system would be given by:

$$W_e = \frac{\varepsilon}{2} \int_{vol} \nabla \varphi \cdot \nabla \varphi \, dV \qquad (2.25)$$

There could be an incremental change in the energy function due to displacement of charges from their mean position. This is given by:

$$\delta W_e = \frac{\varepsilon}{2} \left[\int_{vol} \nabla \delta\varphi \cdot \nabla \delta\varphi \, dV \right] \qquad (2.26)$$

Table 2.1 Various identities based on boundary conditions (courtesy of Crampagne et al. [20])

Geometric configuration	Formula to obtain Green's function	Differential equation	γ	I_s
Electric wall at $x = 0$ and c	$G = \sum_{n=1}^{\infty} G_n(y) \sin \frac{n\pi x}{c}$	$\left[\frac{d^2}{dy^2} - \left(\frac{n\pi}{c}\right)^2\right] G_n(y) = -\frac{2}{c\varepsilon}\delta(y-y_o)\sin\frac{n\pi x_o}{c}$	$\frac{n\pi}{c}$	$\frac{2}{n\pi}\sin\frac{n\pi x_o}{c}$
Electric wall at $x = 0$ and magnetic wall at $x = c$	$G = \sum_{n=0}^{\infty} G_n(y) \sin \frac{(2n+1)\pi x}{2c}$	$\left[\frac{d^2}{dy^2} - \left(\frac{(2n+1)\pi}{2c}\right)^2\right] G_n(y) = -\frac{2}{c\varepsilon}\delta(y-y_o)\sin\frac{(2n+1)\pi x_o}{2c}$	$\frac{(2n+1)\pi}{2c}$	$\frac{4}{(2n+1)\pi}\sin\frac{(2n+1)\pi x_o}{2c}$
Magnetic walls at $x = 0$ and c	$G = \sum_{n=1}^{\infty} G_n(y) \cos \frac{n\pi x}{c}$	$\left[\frac{d^2}{dy^2} - \left(\frac{n\pi}{c}\right)^2\right] G_n(y) = -\frac{2}{c\varepsilon}\delta(y-y_o)\cos\frac{n\pi x_o}{c}$	$\frac{n\pi}{c}$	$\frac{2}{n\pi}\cos\frac{n\pi x_o}{c}$

Fig. 2.5 Lateral view of general microstrip-like interconnect structure. © 1982 IEEE reprinted, with permission, from Bhat and Koul [2]

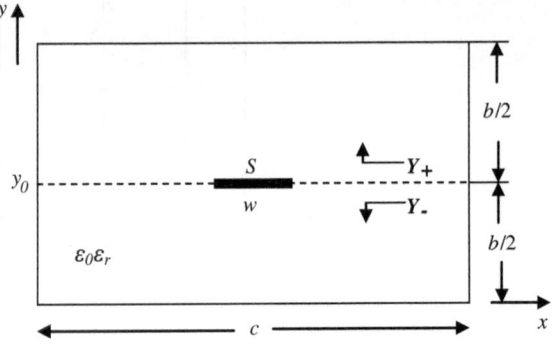

Authors in [1, 14] have elaborated on the upper and lower bounds of this capacitance. The upper bound on capacitance per unit length of the line is given by:

$$C = \frac{\varepsilon}{V_o^2} \iint\limits_{xy-plane} |\nabla_t \phi|^2 dx \, dy = \frac{\varepsilon \iint\limits_{xy-plane} |\nabla_t \phi|^2 dx \, dy}{\left(\int\limits_{S_1}^{S_2} \nabla_t \phi \cdot dl \right)^2} \qquad (2.27)$$

where, V_o is the line integral of $\nabla_t \phi$ from S_1 to S_2. For an approximate value of φ, the calculated value of C will always be greater than the true value.

Similarly, the lower bound on capacitance is given by:

$$\frac{1}{C} = \frac{1}{Q^2} \int\limits_{S_2} \varphi(x,y) \rho(x,y) dl \qquad (2.28)$$

Note that for any trial function $\rho(x_o, y_o)$, the calculated value of $1/C$ is always larger than the true value, which defines the lower bound. We now combine the two techniques to compute the capacitance per unit length for single and coupled transmission line interconnects.

2.2.4 Unified Approach for Calculation of Capacitance Per Unit Length for Single Interconnect Lines

In the earlier section we have derived the variational expression for the capacitance of any two-conductor line having an arbitrary cross-section, as shown in Fig. 2.5. For such an interconnect geometry the charge distribution is given as:

$$\rho(x,y) = f(x)\delta(y - y_o) \qquad (2.29)$$

Substituting this charge distribution function in (2.28), the variational formula for the capacitance per unit length for a multilayer structure with side walls is given as:

$$\frac{1}{C} = \frac{\int\limits_S \int\limits_S G(x,y/x_o,y_o)f(x)dxdx_o}{\left[\int\limits_S f(x)dx\right]^2} \tag{2.30}$$

Here, Green's function for various boundary conditions derived in the previous section as (2.20), (2.22) and (2.24) can be substituted in (2.29). We get the expressions for capacitance for the three cases of boundary conditions as:

Case a

$$C = \frac{\left[\int\limits_S f(x)dx\right]^2}{\sum\limits_{n=1}^{\infty} \frac{2}{n\pi Y}\left[\int\limits_S f(x)\sin\frac{n\pi x}{c}dx\right]^2} \tag{2.31}$$

Case b

$$C = \frac{\left[\int\limits_S f(x)dx\right]^2}{\sum\limits_{n=0}^{\infty} \frac{4}{(2n+1)\pi Y}\left[\int\limits_S f(x)\sin\frac{(2n+1)\pi x}{2c}dx\right]^2} \tag{2.32}$$

Case c

$$C = \frac{\left[\int\limits_S f(x)dx\right]^2}{\sum\limits_{n=1}^{\infty} \frac{2}{n\pi Y}\left[\int\limits_S f(x)\cos\frac{n\pi x}{c}dx\right]^2} \tag{2.33}$$

At any interface of the dielectrics the admittance can be decomposed into two parts: the Y_+ and Y_- representing the admittances above and below the charge plane, respectively. The total admittance at the charge plane is a parallel combination of these two terms and is therefore a summation of the upper and lower admittances as $Y = Y_+ + Y_-$. Using the expression for the input admittance Y_{in} the admittance of a particular section l_j can be computed. The input admittance $Y_{in\,j}$ is given by:

$$Y_{inj} = Y_{cj}\left[\frac{Y_{lj} + Y_{cj}\tanh(\gamma_j l_j)}{Y_{cj} + Y_{lj}\tanh(\gamma_j l_j)}\right] \tag{2.34}$$

where Y_{lj} is the load admittance of the jth section which will be the input admittance $Y_{in\ j+1}$ of the next $(j + 1)$th section. Also, Y_{cj} and γ_j are the characteristic admittance and propagation constant of the jth section.

$$Y_{cj} = \varepsilon_j \tag{2.35}$$

and

$$\begin{aligned} \gamma_j = \gamma &= n\pi/c, &&\text{for cases (a) and (c)} \\ &= (2n + 1)\pi/2c, &&\text{for case (b)} \end{aligned}$$

We now need to calculate the charge distribution $f(x)$ before performing the integration in capacitance formula above. The charge density is concentrated at the edges of the strip. For such a case, the function is given by:

$$\begin{aligned} f(x) &= \left[1 - \{(2/w)|x - c/2|\}^2\right]^{-1} &&(c - w)/2 \leq x \leq (c + w)/2 \\ &= 0 &&\textit{otherwise} \end{aligned} \tag{2.36}$$

Here w is the width of the strip conductor. Authors in [14] report an accurate trial function and is given by (2.37):

$$\begin{aligned} f(x) &= (1/w)[1 + A|(2/w)(x - c/2)|^3], &&(c - w)/2 \leq x \leq (c + w)/2 \\ &= 0 &&\textit{otherwise} \end{aligned} \tag{2.37}$$

Substituting (2.37) in (2.31) and simplifying, the unified formula for capacitance per unit length is:

$$C = \frac{(1 + 0.25A)^2}{\displaystyle\sum_{n\,odd} \left((L_n + AM_n)^2 P_n/Y\right)} \tag{2.38}$$

where

$L_n = \sin(\beta_n w/2)$

$M_n = (2/\beta_n w)^3 \left[3\{(\beta_n w/2)^2 - 2\}\cos(\beta_n w/2) + (\beta_n w/2)\{(\beta_n w/2)^2 - 6\}\sin(\beta_n w/2) + 6\right]$

$P_n = (2/n\pi)(2/\beta_n w)^2$

$\beta_n = n\pi/c$

$$A = -\frac{\displaystyle\sum_{n\,odd} (L_n - 4M_n)L_n P_n/Y}{\displaystyle\sum_{n\,odd} (L_n - 4M_n)M_n P_n/Y}$$

$$\tag{2.39}$$

Note that in (2.38), we only need to evaluate the admittance Y at the charge plane depending on the structure under investigation and the corresponding boundaries present.

Fig. 2.6 Lateral view of
edge-coupled microstrip-like
interconnect structure.
© 1982 IEEE reprinted, with
permission, from Bhat and
Koul. [2]

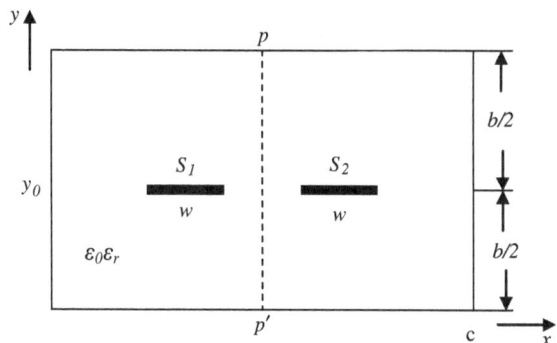

2.2.5 Unified Approach for Calculation of Capacitance for Coupled Interconnect Lines

In case of edge-coupled stripline structure, as shown in Fig. 2.6, the even- and odd-mode capacitances can be obtained by placing a magnetic wall and an electric wall, respectively, at the center of the coupled lines, and by considering half the structure between $x = 0$ and $x = c/2$. The even- and odd-mode charge distribution functions are assumed to be of the form given by:

$$f(x)_{\left(\substack{even\\odd}\right)} = \left(\frac{1}{w}\right)\left[1 + A_{\left(\substack{even\\odd}\right)}\left|\left(\frac{2}{w}\right)(x - (c - s - w)/2)\right|^3\right], \quad for \, ((c - s)/2 - w \le x \le (c - s)/2$$

$$= 0, \qquad\qquad\qquad\qquad\qquad\qquad\qquad\qquad\qquad otherwise.$$

$$(2.40)$$

Applying the transverse transmission line method, the even- and odd-mode Green's function can be expressed as:

$$G(x, y_0/x_0, y_0)_{\left(\substack{even\\odd}\right)} = \sum_{n\left(\substack{even\\odd}\right)} (4/n\pi Y) \sin(\beta_n x) \sin(\beta_n x_0), \qquad (2.41)$$

where

$$\beta_n = n\pi/c. \qquad (2.42)$$

The expression for the admittance Y at the charge plane $y = y_0$ for each coupled line structure is the same as that for the corresponding single line conductor configuration (as discussed in the previous subsection). The variational expression for the capacitance C is given by:

$$C_{\left(\substack{even\\odd}\right)} = \frac{\left(1 + 0.25A_{\left(\substack{even\\odd}\right)}\right)^2}{\sum_{n\left(\substack{odd\\even}\right)} \left(\left(L_n + M_n A_{\left(\substack{even\\odd}\right)}\right)^2 P_n/Y\right)}, \qquad (2.43)$$

where

$$L_n = \sin(\beta_n w/2)\sin\left\{\beta_n\left(\frac{c-s-w}{2}\right)\right\},$$

$$M_n = (2/\beta_n w)^3 \sin\left\{\beta_n\left(\frac{c-s-w}{2}\right)\right\}\left[\begin{array}{l}3\left\{(\beta_n w/2)^2 - 2\right\}\cos(\beta_n w/2)\\[2mm]+(\beta_n w/2)\left\{(\beta_n w/2)^2 - 6\right\}\sin(\beta_n w/2) + 6\end{array}\right],$$

$$P_n = (4/n\pi)(2/\beta_n w)^2, \; and$$

$$A_{odd}^{even} = -\frac{\displaystyle\sum_{n_{even}^{odd}}(L_n - 4M_n)L_n P_n/Y}{\displaystyle\sum_{n_{even}^{odd}}(L_n - 4M_n)M_n P_n/Y}.$$

$$(2.44)$$

2.3 Concluding Remarks

This chapter presents detailed derivation of the unified approach. We present a qualitative comparison of some of the widely used analytical techniques for analysis of transmission line interconnects. In that we feel that the unified approach is best suited for parameter extraction, computation of propagation constants and CAD programs. It is simple yet accurate with less computational resources. Chip–chip interconnects are essentially planar transmission lines and this technique can be used for analysis of single edge- and broadside-coupled interconnects.

The unified approach essentially combines transverse transmission line technique for computation of Green's function with the variational method for capacitance calculations. Nowadays, high-speed interconnects have a complex layout with presence or absence of ground. This technique can be very efficiently used to address these modified boundary conditions. In the next chapter we shall illustrate the application of this technique to such interconnect structures. In that single and coupled interconnects with modified geometrical configurations will be analyzed. These modified geometries suggest recalculation of the admittance parameters based on the appropriate boundary conditions. The authors strongly recommend further reading of the unified approach and other techniques proposed by authors in [1, 2, 7, 20] for greater understanding of analytical approaches that can be used in developing compact interconnect models.

References

1. R.E. Collin, *Field Theory of Guided Waves* (McGraw-Hill, New York, 1960)
2. B. Bhat, S.K. Koul, Unified approach to solve a class of strip and microstrip-like transmission lines. IEEE Trans. Microw. Theory Tech. **82**(5), 679–686 (1982)

3. L.A. Pipes, *Applied Mathematics for Engineers and Physicists* (McGraw-Hill, New York, 1958)
4. S.B. Cohn, Shielded coupled-strip transmission line. IRE Trans. Microw. Theory Tech. **3**(5), 29–38 (1955)
5. S.B. Cohn, Characteristic impedances of broadside-coupled strip transmission lines. IRE Trans. Microw. Theory Tech. **8**(6), 633–637 (1960)
6. J.P. Shelton, Impedances of offset parallel-coupled strip transmission lines. IEEE Trans. Microw. Theory Tech. **14**(1), 7–14 (1966)
7. H.E. Green, The numerical solution of some important transmission line problems. IEEE Trans. Microw. Theory Tech. **13**(5), 676–692 (1965)
8. M.V. Schneider, Computation of impedance and attenuation of tem lines by finite-difference methods. IEEE Trans. Microw. Theory Tech. **13**(6), 793–800 (1965)
9. J.G. Yook, N.I. Dib, L.P.B. Raheti, Characterization of high frequency interconnects using finite difference time domain and finite element method. IEEE Trans. Microw. Theory Tech. **42**(9), 1727–1736 (1994)
10. O.S. Rosales, D. Suster, Finite-difference computation of the characteristic impedance of unbounded striplines and microstrip lines, in *Proceedings of the 1st IEEE International Caracas Conference on Devices, Circuits and Systems*, (1995) p. 323–327
11. B.N. Das, K.V.S.V.R. Prasad, A generalized formulation of electromagnetically-coupled striplines. IEEE Trans. Microw. Theory Tech. **32**(11), 1427–1433 (1984)
12. E. Yamashita, Variational method for the analysis of microstrip-like transmission lines. IEEE Trans. Microw. Theory Tech. **16**(8), 529–535 (1968)
13. E. Yamashita, R. Mitra, Variational method for the analysis of microstrip lines. IEEE Trans. Microw. Theory Tech. **16**(4), 251–256 (1968)
14. B. Bhat, S.K. Koul, *Stripline-like Transmission Lines for Microwave Integrated Circuits* (Wiley, New York, 1989)
15. R.E. Diaz, The discrete variational conformal technique for the calculation of strip transmission line parameters. IEEE Trans. Microw. Theory Tech. **34**(6), 714–722 (1986)
16. D.W. Kammler, Calculation of characteristic admittances and coupling coefficients for strip transmission lines. IEEE Trans. Microw. Theory Tech. **16**(11), 925–937 (1968)
17. T. Itoh, R. Mittra, A technique for computing dispersion characteristics of shielded microstrip lines. IEEE Trans. Microw. Theory Tech. **22**(10), 896–898 (1974)
18. J.B. Davies, D. Mirshekar-Syahkaf, Spectral domain solution of arbitrary transmission line with multilayer substrate. IEEE Trans. Microw. Theory Tech. **25**(2), 143–146 (1977)
19. J.I. Smith, The even- and odd-mode capacitance parameters for coupled lines in suspended substrate. IEEE Trans. Microw. Theory Tech. **19**(5), 424–431 (1971)
20. R. Crampagne, M. Ahmadpanah, J.L. Guiraud, A simple method for determining the green's function for a large class of MIC lines having multilayered dielectric substrates. IEEE Trans. Microw. Theory Tech. **26**(2), 82–87 (1978)

Chapter 3
Compact Models for Novel Interconnects Using Unified Approach

In the last chapter we presented a qualitative summary of some of the commonly used modeling approaches. We concluded that the unified approach as presented by authors in [1–5] provides an efficient and accurate modeling technique. However, the previous publications by the same authors limit the use of this technique to MIC and MMIC applications only. Since chip–chip interconnects resemble planar transmission lines, the unified approach can be used to effectively model these high-speed interconnects as well. Having laid the foundation for the use of the unified approach in the previous chapter, we now present its applicability in analyzing high-speed interconnects. In particular, we present a composite parameter extraction algorithm followed by models for transient analysis of these interconnects. The application of unified approach for more standard structures such as the microstrip and stripline-like interconnects is already explained in the available literature and is therefore not covered in this chapter. Rather we present the analysis of complex yet practical interconnect geometries that are used in Printed Circuit Board (PCB) and multichip module (MCM) applications nowadays.

In conjunction with the requirement of miniaturization of devices, design engineers do not have the luxury of indulging in repeated full-wave simulation of the design, specifically high-speed layout design. Thus, it is now becoming imperative to use circuit model [6, 7] at the beginning followed by final verification, done using full-wave simulators; the emphasis is on reduced time-frame while maintaining reasonable accuracy. Two more emergent needs are worth mentioning; firstly due to miniaturization efforts, component density on a mixed-signal board has gone up manifolds and secondly the conventional methods of PCB trace design with neat ground layers are no longer feasible in most of the design efforts [8, 9]. At the Industrial, Scientific and Medical (ISM) frequency bands of 2.4/5.8 GHz, the interconnect lines from one module to another module are not electrically short and thus the performance of such interconnects is critically dependent on the proximity of ground trace (or absence of such) in both multilayer and coplanar configurations. There is need for routing of PCB traces,

R. Sharma and T. Chakravarty, *Compact Models and Measurement Techniques for High-Speed Interconnects*, SpringerBriefs in Electrical and Computer Engineering, DOI: 10.1007/978-1-4614-1071-3_3, © The Author(s) 2012

where PCB trace in one layer sees small windows of aperture in the ground plane in the adjacent layer [10]. Similarly coplanar ground proximity may not run uniformly throughout the entire run of the interconnect line. Also, the typical problem of signal integrity issues in dense layout continues to pose problems [11–13, 14].

From the above discussion, it is felt that a unified approach to compute quickly and accurately (even though quasistatic) the various line parameters such as characteristic impedance, delay, cross-talk for an arbitrary placement of ground trace relative to the interconnect line are needed. It is definitely desirable that the approach should offer physical insight into the range of problems addressed so that much of the design cycle time can be effectively reduced through intuitive, yet rational, routing scheme of the interconnect and associated ground trace.

In this chapter we present the analysis of these novel interconnects using the unified approach. It is felt that this modeling approach can be used effectively for a range of interconnect geometries over multiple layers of dielectrics. In the following section, we explain how the distributed line parameters are extracted using this approach. Once the per unit length parameters are known, we can obtain line impedance and delay time constants using transmission line formulas. Next we present an efficient algorithm that can be used for CAD applications based on the unified approach. In the subsequent section, we study some of the novel interconnects to demonstrate the applicability of this approach and are presented as case studies. At the end of the chapter, the readers should have a clear idea of this approach and should be able to apply it in various interconnect geometries of interest.

3.1 Computation of Distributed Line Parameters, Impedance and Delay Constants

The unified approach-based formula for capacitance per unit length for single and coupled interconnect lines has been derived in the previous chapter and is given as (2.38), (2.39), (2.43) and (2.44). Note that the admittance in these expressions was computed using the transverse transmission line technique. This is a generic formula and can be applied to any multilayered planar transmission line with 'Y' as the only parameter to be determined. The capacitance expression represents the overall line capacitance per unit length, and is a summation of various field components including the overlap ($C_{overlap}$), lateral ($C_{lateral}$) and fringe capacitances (C_{fringe}), as shown in Fig. 3.1.

The inductance and capacitance of multiconductor transmission lines are related and is given by [3, 15] as:

$$L = 1 \Big/ (v^a)^2 C^a \tag{3.1}$$

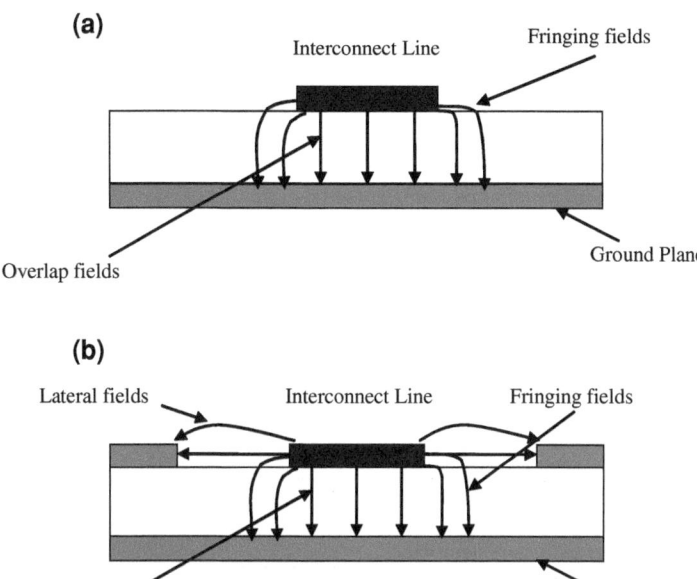

Fig. 3.1 **a** Field distribution (single interconnect line), **b** Field distribution (interconnect line with coplanar ground track)

Here, L is the distributed inductance of the line, C^a is the line capacitance when all dielectric constants are equal to 1 and v^a is the velocity of propagation in air. Calculation of R is relatively simple as it is function of the interconnect geometry and conductivity and is independent of adjoining conductors or modified ground planes. It is given by:

$$R = \frac{1}{\sigma t w} \tag{3.2}$$

However, as signal frequency increases the skin depth $(\delta = 1/\sqrt{\pi f \mu \sigma})$ in the conductor decreases thereby increasing the line resistance [16]. Skin effect cannot be ignored for frequencies above a few GHz since the depth of current penetrating into the metal conductor is comparable to the conductor width. Skin effect at high frequencies along with proximity effect due to inductive coupling of neighboring wires may increase the line resistance significantly. In such a scenario, Cao et al. [17] propose the frequency-dependent $R_{ac}(f)$ to be used for more accurate resistance extraction.

We are now in possession of the distributed RLC parameters for any planar interconnect structure based on the unified expression for line capacitance and AC-resistance formula. The unified formula for capacitance can also be used to compute the line impedance. The characteristic impedance Z_0 can now be computed as $Z_0 = 1/v^a\sqrt{C \cdot C^a}$ [3]. Here, C is the capacitance per unit length of the

Table 3.1 Admittance computation algorithm

Input: Given interconnect geometry, including width (w), thickness (t), substrate permittivity (ε_r)
Output: Admittance at the charge plane $y = y_0$

BEGIN

1. Identify the charge plane $y = y_0$, the interface between two layers, at which the interconnect line of thickness 't' is situated between $y = y_0$ and $y = y_0 \pm t$
2. Determine the boundary conditions on the rectangular walls, do the following
a. Compute Green's function using identities $\sin\frac{n\pi x}{c}$, $\sin\frac{(2n+1)\pi x}{2c}$ and $\cos\frac{n\pi x}{c}$ being orthogonal between $x = 0$ and c
3. Develop the equivalent transmission line circuit [1] using the equivalencies $Y_{cj} \equiv \varepsilon_j$ and $\gamma_j = n\pi/c$ or $(2n + 1)\pi/2c$. Compute the input and load admittances for all dielectric layers above and below the charge plane
4. Obtain overall admittances for the regions above and below the charge plane as Y_+ and Y_-, respectively
5. Compute total admittance Y seen at the charge plane as $Y = Y_- + Y_+$

END

Table 3.2 Parameter extraction algorithm

Input: Given interconnect geometry, including width (w), thickness (t), substrate permittivity (ε_r), and admittance (Y) at the charge plane $y = y_0$
Output: SPICE-compatible RLC parameters per unit length

BEGIN

1. Compute line capacitance 'C' using (2.38–2.44) and 'Y' from Table 3.1
2. Compute line inductance 'L' using (3.1) by equating all dielectric constants to 1 and computing C^a
3. Determine R using (3.2);
a. Compute skin depth 'δ' according to extraction frequency 'f'. Calculate $R_{ac}(f)$
4. Compute characteristic impedance 'Z' from steps 1 and 2 above

END

structure, C^a is the capacitance per unit length of the structure with all dielectrics replaced by air and v^a is the velocity of propagation in air. The above equations can be used to obtain the Simulation Program with Integrated Circuit Emphasis (SPICE)-compatible equivalent parameters. The above discussion leads to a detailed admittance computation and parameter extraction algorithm, which is summarized in Tables 3.1 and 3.2.

Once the R, L, C parameters are extracted for a given structure, corresponding damping factor can be computed. The damping factor ζ for any second-order system is given by:

$$\zeta = \frac{RC}{2\sqrt{LC}} \tag{3.3}$$

We can now use the above parameter extraction formulation to compute time-domain response for the interconnect line. In the previously published literature several closed-form expressions are derived for computation of delay parameters.

We would like to highlight the work presented by authors in [18, 19], in that they proposed that the second-order approximation of the transfer function of an RLC interconnect can be used to determine the output time-domain signal for any arbitrary input. For a step input and a supply voltage of V_{DD}, the output time domain response derived from the second-order approximation is given by:

$$S(t) = V_{DD} + \frac{V_{DD}}{2\sqrt{\zeta^2 - 1}} \left[\frac{\exp\left[\omega_n t \left(-\sqrt{\zeta^2 - 1}\right)\right]}{-\zeta + \sqrt{\zeta^2 - 1}} - \frac{\exp\left[\omega_n t \left(-\zeta - \sqrt{\zeta^2 - 1}\right)\right]}{-\zeta - \sqrt{\zeta^2 - 1}} \right]$$

(3.4)

where, ζ is given by (3.3) while $\omega_n = 1/\sqrt{LC}$.

Thus, from (2.38), (2.43), (3.1), (3.2) we can now obtain the time-domain response of a high-speed interconnect line using second-order approximation. For an arbitrary input 50% delay time (t_d), rise time (t_r), maximum and minimum signal overshoots (%O) and their corresponding overshoot time (t_o) and settling time (t_s) can be computed considering both monotonic and non-monotonic signal responses. These closed form delay expressions are reproduced as (3.5–3.9) from [19] and simplified for the sake of continuity. However, for details readers are referred to the previously published literature on RLC interconnects including [18, 19].

$$t_d = \left(1.047 \ \exp\left(\frac{\zeta}{0.85}\right) + 1.39 \ \zeta \right)$$

(3.5)

$$t_r = \left(6.017 \ \exp\left(\zeta^{1.35}/0.4\right) - 5 \ \exp\left(\zeta^{1.25}/0.64\right) + 4.39 \ \zeta \right)/\omega_n$$

(3.6)

$$\%O = (-1)^{n+1}.100 \ \exp\left(-\frac{n\pi\zeta}{\sqrt{1 - \zeta^2}}\right), \quad n = 1, 2, 3, \ldots$$

(3.7)

$$t_0 = \frac{n\pi}{\omega_n\sqrt{1 - \zeta^2}}$$

(3.8)

$$t_S = \frac{2.3}{\zeta\omega_n}$$

(3.9)

Based on the above discussion we now have a general analytical tool that can be either programed as post-extraction computer software or combined into an existing extraction tool. The above-mentioned analytical approach and algorithm [20] retains the same accuracy and simplicity as that of unified approach with minimum investment of computational resources unlike 2D/3D field solvers. The proposed unified approach provides effective design space exploration and hence is useful for optimization, analysis and synthesis of high-speed interconnects.

Fig. 3.2 Standard microstrip
line

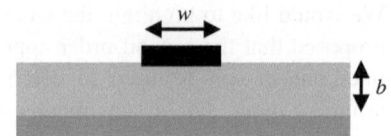

3.2 Case Study: Application of Unified Approach

In this section we present some case studies; in that we analyze novel interconnect
topologies that can be used in PCB and MCM chip–chip connections. While a
limited number of single and coupled interconnects are analyzed, the analytical
approach is general and can be used to analyze multilayer transmission line
interconnects for all practical purposes.

3.2.1 Analysis of Modified Single Interconnect Lines

The application of unified approach to analyze microstrip lines and striplines is
clearly explained in the previous literature. Therefore we shall not discuss those
here, rather we shall show how this approach can be used to analyze interconnect
lines that represent modified microstrip configuration with altered boundary conditions. Nowadays conventional methods of PCB trace design with neat ground
layers are no longer feasible in most of the design efforts. As mentioned earlier, the
performance of such interconnects is critically dependent on the proximity of
ground trace (or absence of such) in both multilayer and co-planar configurations.
Our proposed compact models for high-speed interconnects can come to our
rescue as it gives the designer physical insights and aid in design space exploration
without the use of time-consuming simulations. It is felt that using the unified
approach, we can compute quickly and accurately (even though quasi-static), the
various line parameters such as characteristic impedance, delay, cross-talk for an
arbitrary placement of ground trace relative to the interconnect line can be
extremely useful.

Figure 3.2 represents a standard microstrip line and Fig. 3.3 shows the various
possible modifications to the microstrip configuration. The interconnect line has a
width 'w', and a substrate height 'b'. Any coplanar ground track has an edge–edge
distance 'd' while any aperture in the ground below has an aperture width 'w_s'.
These structures exhibit alteration in the position of ground plane resulting in
modified boundary conditions. This can complicate analysis as well as parasitic
extraction in such interconnects. We now demonstrate how the proposed unified
model can compute the admittance of multilayered planar transmission lines with
several variations in ground placement. Once the admittance parameter is computed, the equivalent circuit parameters can be extracted based on the formulation
given in the previous section.

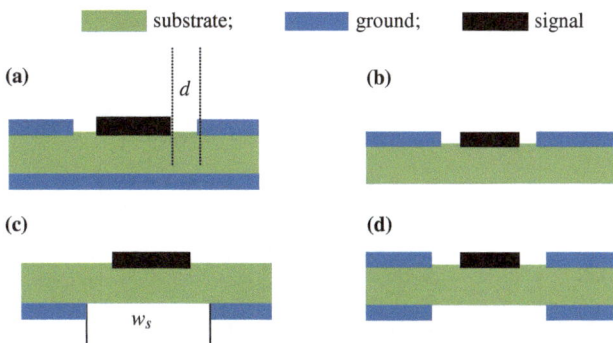

Fig. 3.3 Modified interconnect lines with altered ground placement. **a** Microstrip with coplanar ground. **b** Coplanar interconnect line. **c** Microstrip with ground plane aperture. **d** Microstrip with ground plane aperture and coplanar ground

It is seen that the presence or absence of ground tracks adjacent to signal lines may alter the electrical characteristics of the interconnect line itself. This is due to the modified boundary conditions offered by the altered ground patterning. Traditionally, the impedance of the microstrip-like line depends on the width of the line, the permittivity and the height of the substrate. In the modified interconnect geometry, signal lines are flanked by grounded guard tracks and the characteristic impedance now becomes a function of the spacing between the interconnect line and the coplanar ground tracks and/or the aperture width. In our computation we assume an arbitrary charge plane $y = y_o$ at the interface of the dielectric, where the signal line is placed. Thus the entire structure can be divided into two regions with two distinct admittances Y_+ and Y_- representing the respective admittances of the upper and lower regions. As a case in point, we analyze a microstrip line with coplanar ground which is elaborated graphically in Fig. 3.4. It is a two layer structure with the bottom layer given by dielectric 1 (ε_1, b_1) and the top layer given by dielectric 2 (ε_2, b_2) bounded by electric shorts on all four sides. We are interested to compute the admittance at the interface of these two layers on the charge plane.

For such computation, we use the standard transmission line formulation given by:

$$Y_{in,j} = \varepsilon_j \left[\frac{Y_{ij} + Y_{cj} \tanh(\gamma_j l_j)}{Y_{cj} + Y_{ij} \tanh(\gamma_j l_j)} \right] \tag{3.10}$$

If we now consider the admittance (Y_-) seen at the charge plane due to lower region, we obtain the result as:

$$Y_{Lower} = \varepsilon_0 \varepsilon_1 \coth(\beta_n b_1)$$
$$\beta_n = n\pi/c \tag{3.11}$$
$$c = 2.d + w,$$

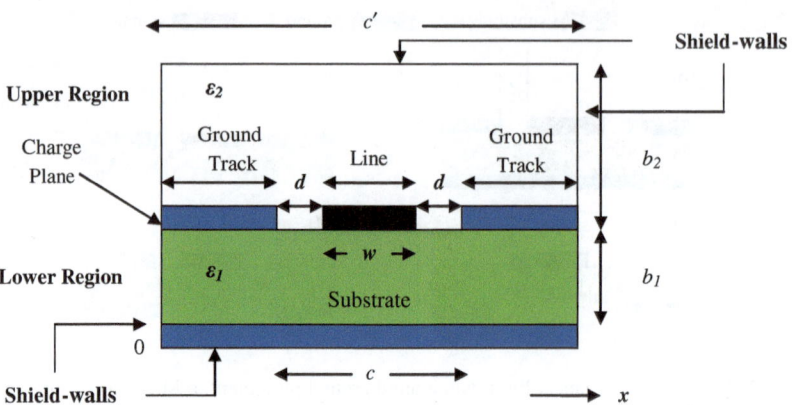

Fig. 3.4 Cross-section of a microstrip line with coplanar ground with distinct upper and lower regions

this is obtained since $Y_{ij} = \infty$ (electric short at the bottom). For this case, ε_1 and b_1 are the permittivity and the height of the dielectric layer, respectively, and is computed for odd values of n excluding $n = 0$. The distance c is shown by dotted lines. Similarly $Y_{in,j}$ is iterated over each section of the transmission line to determine the individual admittances of the lower and upper region and summing them to obtain Y at the charge plane. The admittance of the upper region is given by:

$$Y_{Upper} = \varepsilon_0\varepsilon_2 \coth(\beta_n b_2)$$
$$\beta_n = n\pi/c'$$
$$c' \gg w$$
$$b_2 \gg b_1, \tag{3.12}$$

where, ε_2 and b_2 are the permittivity and the height of the dielectric layer (upper region), respectively, and is computed even for values of n excluding $n = 0$. Here c' is a variable distance and is kept much greater than the line width 'w'. Substituting (3.11) and (3.12) in (2.38), we compute the line capacitance for these two regions, C_{Lower} and C_{Upper}, respectively. The total capacitance will now be the summation of C_{Lower} and C_{Upper}.

The capacitance formula given by (2.38) is applicable to any single conductor stripline-like transmission line interconnects with one or more dielectric layers. If the interconnect has a small but finite thickness 't', (2.38) can still be used by replacing Y in (3.11) and (3.12) by Y/h (β_n, t), as reported in [1, 2]. The expression for h (β_n, t) for the structure considered is given by

$$h(\beta_n, t) = \frac{1}{2}\left[1 + \frac{\sinh\{\beta_n(b_2 - t)\}}{\sinh\{\beta_n b_2\}}\right]. \tag{3.13}$$

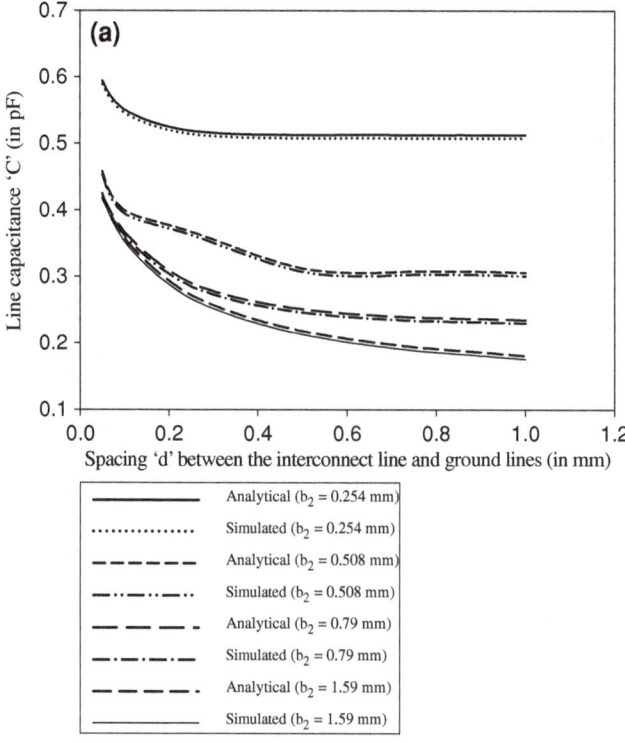

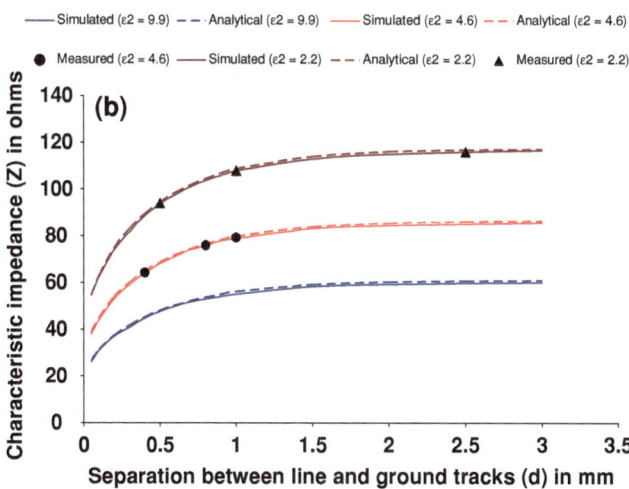

Fig. 3.5 **a** Simulated and analytical data for line capacitance ($w = 1$ mm, $\varepsilon_2 = 4.6$), **b** Simulated, predicted and measured characteristic impedance ($w = 1$ mm and $b_2 = 1.59$ mm)

Thus an interconnect line of finite thickness can also be analyzed using the unified model. The set of equations given in Sect. 3.1 can now be used to compute inductance, impedance and delay constants. The method of computing admittance using the unified approach is efficient yet accurate. Comparison of the analytical results for capacitance and line impedance with 3D field solver and measurements shows good accuracy with less than 5% error. This is shown in Fig. 3.5.

The absence of ground below the interconnect line can be attributed to a partly etched ground plane or a defected ground structure. For such a case, consider an interconnect structure shown in Fig. 3.6, with an electric wall separation 'c' forming an enclosed cavity with part opening due to an aperture in the ground plane. Figure 3.6 shows a three-layer vertical structure represented by dielectric layers 1 (ε_1, b_1), 2 (ε_2, b_2) and 3 (ε_3, b_3), respectively where ε is the permittivity and b is the thickness of the dielectric layers. The geometry of the proposed structure that is an interconnect line with symmetrical ground plane aperture is represented by an interconnect line at the interface between regions 2 and 3 and the partial ground plane projection at the interface between regions 1 and 2. For a microstrip structure, region 1 has no role since the interface between regions 1 and 2 contains a ground plane from end to end wall. On the other hand, for suspended stripline structure, all the three regions enclosed by the shielding walls play a role in the computation of parameters like characteristic impedance. In this study, the structure proposed is a microstrip line with ground plane aperture. Therefore this problem is approached by considering both the cases of microstrip and suspended stripline structures. From Fig. 3.6 one is able to see that the proposed structure is nonhomogeneous. This is due to the fact that the ground extends to a certain distance below the substrate from the end walls with partial opening, indicated as ground plane aperture symmetrically located below the transmission line on the charge plane.

Based on this reasoning we divide region 2 into three vertical profiles as shown in Fig. 3.7 namely I, II and III, respectively. Individual admittances of the three vertical profiles computed on the charge plane that is the interface between regions 2 and 3 are parallel to each other. For computation of characteristic impedance of the line, first, we consider the regions 2 and 3 with finite thickness (b_2 and b_3) and permittivity (ε_2 and ε_3), as shown. The admittance measured on the charge plane (interconnect line) due to these two regions are denoted as Y_{Lower} and Y_{Upper}, respectively. While region 3 is homogenous, region 2 is not. Thus the admittance represented by Y_{Lower} is a parallel combination of admittances on the charge plane due to the three vertical profiles. There are three possible boundary conditions that should be considered in this analysis:

- *Case a*: electric walls at $x = 0$ and c

$$\beta_n = \frac{n\pi}{c}$$

- *Case b*: magnetic walls at $x = (c - w_s)/2$ and $(c + w_s)/2$

Fig. 3.6 Cross-section of the microstrip with ground plane aperture

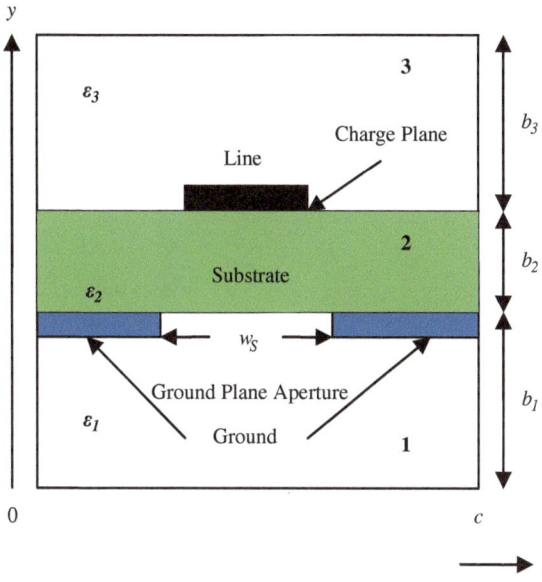

Fig. 3.7 Vertical profiles in region 2

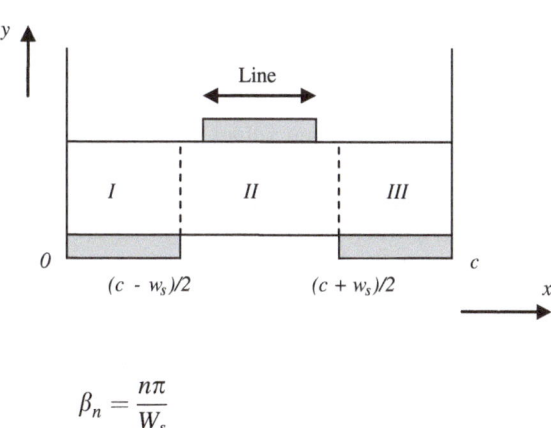

$$\beta_n = \frac{n\pi}{W_s}$$

- *Case c*: electric wall at $x = 0$ and magnetic wall at $x = (c - w_s)/2$

$$\beta_n = \frac{(2n + 1)\pi}{2g}$$

where, $n = 1, 3, 5 \ldots \infty$ and $g = (c - w_s)/2$

Region 3 is a homogeneous medium with electric walls at $x = 0$ and c and $\varepsilon_3 = 1$ (air). The admittance of this region is given by:

$$Y_{Upper} = \varepsilon_0 \varepsilon_3 \coth(\beta_n b_3) \tag{3.14}$$

where, β_n is defined by *Case a* as above. To obtain microstrip-like structure, we have considered open structure at the top ($b_3 \gg b_2$). The admittance of region 2 is given by:

$$Y_{Lower} = Y_I + Y_{II} + Y_{III} \tag{3.15}$$

where,

$$Y_{II} = \varepsilon_0 \varepsilon_2 \tanh(\alpha_n b_2) \tag{3.16}$$

$$Y_I = Y_{III} = \varepsilon_0 \varepsilon_2 \coth(\gamma_m b_2) \tag{3.17}$$

where,

$$\alpha_n = \frac{n\pi}{W_S} \text{ and } \gamma_m = \frac{m\pi}{g} \qquad where\, n = 1, 3 \ldots, \; m = 2, 4 \ldots and\, g = (c - W_s)/2$$

The total admittance of the interconnect structure with the ground plane aperture is given by:

$$Y = Y_{Upper} + Y_{Lower} \tag{3.18}$$

Comparison between analytical results with full-wave simulated data is shown in Fig. 3.8. Again for the proposed models the characteristic impedance closely matches to the simulated values.

Readers by now should have a clear idea of computing admittance using the unified approach based on the above two examples. Thus for a range of modified interconnect structures shown in Fig. 3.3 the admittance can be computed and is summarized in Table 3.3.

Parameter extraction time comparison for interconnects given in Fig. 3.3 is presented in Table 3.4. The tests are run on a Core 2 Quad 2.66 GHz CPU with 4 GB memory. The extraction time using the proposed model is greatly reduced. The reduction may be even more for larger and more complex interconnect systems. Extraction of parasitics is a tedious job and requires either time-consuming simulations or heuristic approaches. However our proposed approach tries to overcome the limitations posed by existing methods, which makes it suitable in CAD-oriented design tools.

3.2.2 Analysis of Modified Coupled Interconnect Lines

We now show the application of the unified approach in analyzing coupled inter-connect lines. The previous literature has already elaborated the application of this approach for analyzing edge-coupled and broadside-coupled lines. We extend the discussion here and consider typical coupled interconnect geometries used in practical chip–chip connections. Coupled noise between interconnect lines is a

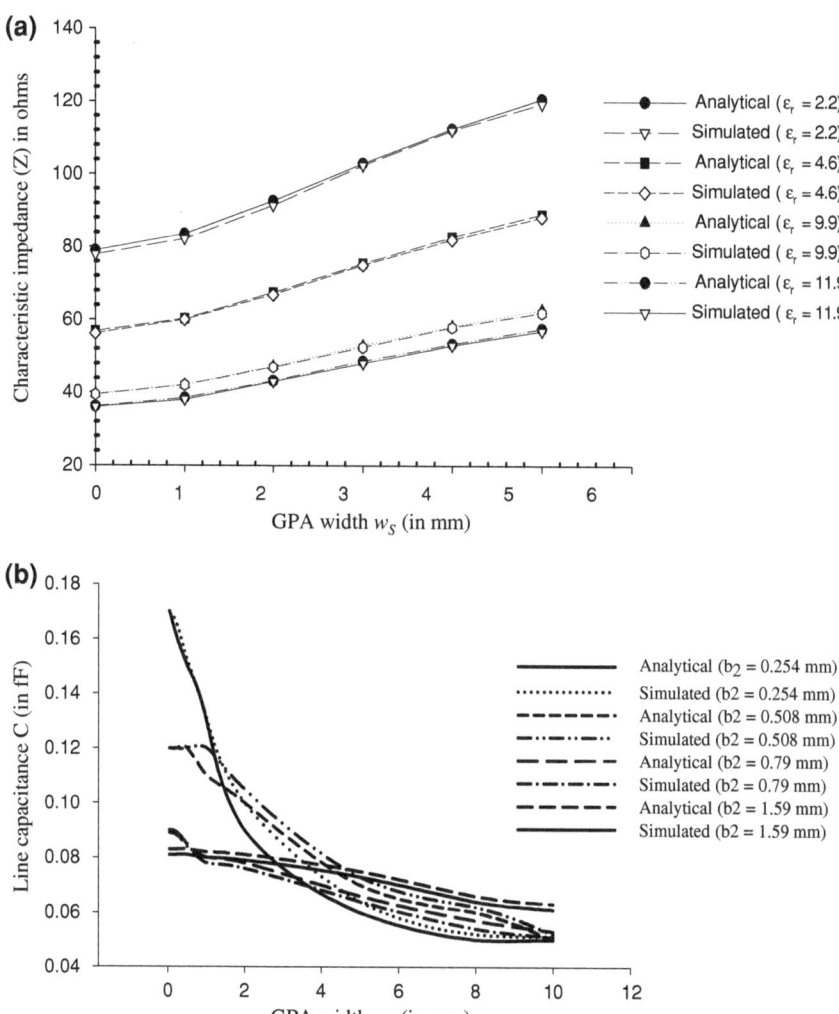

Fig. 3.8 **a** Simulated and analytical characteristic impedance, ($w = 0.76$ mm, and $b_2 = 0.508$ mm), **b** Simulated and analytical line capacitance, ($w = 1$ mm, and $\varepsilon_2 = 4.6$)

potential cause of failure in high-speed digital systems [21–28]. Grounded PCB tracks are often used for reduction in crosstalk in a variety of routing topologies and mixed-signal systems. In case of coupled lines, it is imperative for a designer to verify if a given routing topology will lead to logic failures due to coupled noise. Such verification is typically done using capacitive charge sharing models or exhaustive simulations. The simulation models are pessimistic and time-consuming, so there is a need for simpler methods with better accuracy. Generally coupling parameters between coupled interconnects are computed by calculating the

Table 3.3 Admittance $(Y = Y_- + Y_+)$ for interconnect shown in Fig. 3.3

Fig. 3.3a	Fig. 3.3b

Fig. 3.3a

$$Y_{-,n} = \varepsilon_0 \varepsilon_1 \coth(\beta_n b_1)$$

$$\beta_n = n\pi/c, c = 2.d + w, n = 1,3,5,\ldots$$

$$Y_{+,n} = \varepsilon_0 \varepsilon_2 \coth(\beta_n b_2)$$

$$\beta_n = n\pi/c', c' \gg w, b_2 \gg b_1, n = 2,4,6,\ldots$$

Fig. 3.3b

$$Y_{-,n} = \varepsilon_0 \varepsilon_2 \left(\frac{\coth\left(\beta_n \frac{b_1+b_2}{2}\right) \coth(\beta_n b_2) + \varepsilon_2}{\varepsilon_2 \coth(\beta_n b_2) + \coth\left(\beta_n \frac{b_1+b_2}{2}\right)} \right)$$

$$\beta_n = n\pi/c, c = 2d + w, n = 1,3,5,\ldots$$

$$Y_{+,n} = \varepsilon_0 \varepsilon_3 \coth\left(\beta_n \left(\frac{b_1+b_3}{2}\right) \right)$$

$$\beta_n = n\pi/c', c' \gg w, b_1 = b_3 \gg b_2, n = 2,4,6,\ldots$$

Fig. 3.3c

$$Y_{-,n} = Y_I + Y_{II} + Y_{III}$$

where

$$Y_I = Y_{III} = \varepsilon_0 \varepsilon_2 \coth(\gamma_m b_2)$$

$$Y_{II} = \varepsilon_0 \varepsilon_2 \tanh(\alpha_n b_2)$$

$$\alpha_n = \frac{n\pi}{W_S} \text{ and } \gamma_m = \frac{m\pi}{g}$$

$$n = 1,3,5,\ldots, m = 2,4,6,\ldots, g = (c - W_s)/2$$

$$Y_{+,n} = \varepsilon_0 \varepsilon_3 \coth(\beta_n b_3)$$

$$\beta_n = \frac{n\pi}{c}, b_3 \gg b_2, n = 2,4,6,\ldots$$

Fig. 3.3d

$$Y_{-,n} = Y_I + Y_{II} + Y_{III}$$

where

$$Y_I = Y_{III} = \varepsilon_0 \varepsilon_2 \coth(\gamma_m b_2)$$

$$Y_{II} = \varepsilon_0 \varepsilon_2 \tanh(\alpha_n b_2)$$

$$\alpha_n = \frac{n\pi}{W_S} \text{ and } \gamma_m = \frac{m\pi}{g}$$

$$n = 1,3,5,\ldots, m = 2,4,6,\ldots, g = (c - W_s)/2$$

$$Y_{+,n} = \varepsilon_0 \varepsilon_3 \coth(\beta_n b_3)$$

$$\beta_n = \frac{n\pi}{c}, c = 2d + w, b_3 \gg b_2, n = 2,4,6,\ldots$$

Table 3.4 Parameter extraction time using our proposed algorithm

Interconnect structure	Extraction time (in seconds)
Microstrip with coplanar ground	24
Coplanar interconnect line	25
Microstrip with GPA	31
Microstrip with ground plane aperture and coplanar ground	28

Fig. 3.9 a Coupled interconnect lines with intermediate shield insertion, **b** Differential pair with intermediate shield insertion

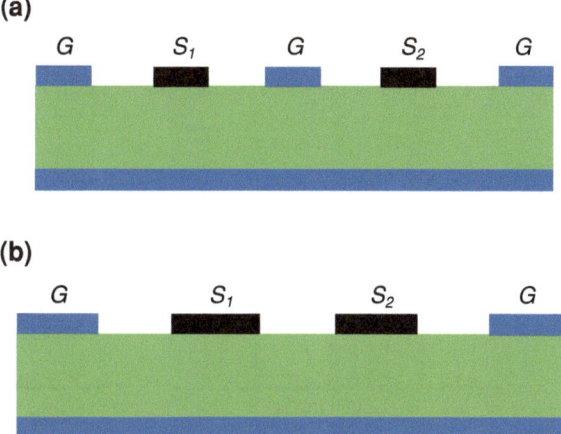

even- and odd-impedances. The use of ground tracks is by far the best remedy for crosstalk-related problems. However the placement of ground tracks adjacent to the signal interconnects imposes modified boundary conditions that require recalculation of the admittance parameters. Figure 3.9 depicts some of these modified cases where regular edge-coupled lines are altered due to the presence of ground or shield lines. Here 'S' represents signal lines while 'G' represents ground lines.

Figure 3.10 shows the lateral view of a multilayered edge-coupled transmission structure with a rectangular shield enclosure. b_2, ε_2 and b_3, ε_3 are the respective heights and permittivity of the regions below and above the charge plane. The separation between the two interconnect lines (line 1 and line 2) is 's' and that between the interconnect lines and outer ground tracks is 'd' and inner ground track is 'd_1'. pp' is an imaginary plane along which the structure is symmetrically placed. $Y+$ and $Y-$ represent the admittances of the regions above and below the charge plane, respectively. The placement of ground tracks on either side of the interconnect line proposes modified boundary conditions. This means that the admittances should be recalculated.

The capacitance formula for coupled interconnect lines is derived in (2.43) and discussed in Chap. 2. The only parameter that needs to be computed in the above formula is the admittance Y of the structure at the charge plane $y = y_0$. The

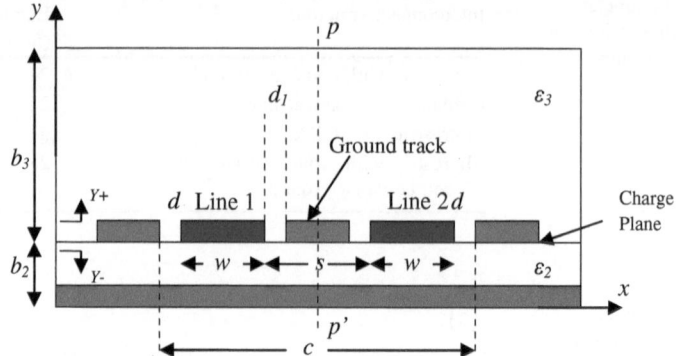

Fig. 3.10 Edge-coupled transmission line structure with intermediate ground track

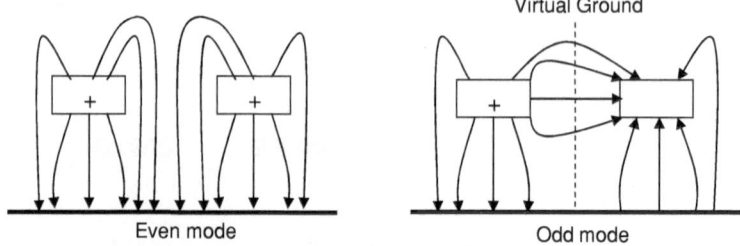

Fig. 3.11 Even- and odd-mode electric field lines

placement of a ground track between the two signal carrying conductors alters the method of computing the admittance on the charge plane from that for conventional microstrip line coupler. This can be explained by observing the field lines for the two modes, namely, odd-and even-modes [29].

Figure 3.11 shows the electric field lines for the two modes necessary for computation of the coupling factor. From Fig. 3.11, it is evident that the odd-mode capacitance is computed by considering an electric wall (virtual ground) between the two conductors. Therefore, a physical placement of a ground between the two conductors does not in anyway alter the computational method. However, for the computation of the even-mode capacitance, we generally consider a magnetic wall between the two conductors. In the present case, however, there exists a physical placement of an electric wall between the two conductors. This ground track extends from the ground plane below the substrate to the charge plane only. Therefore, the admittance presented by the lines now depends on the lateral dimension of wall-to-wall spacing which in this case is given by $g = 2(d + w + d_1)$. Since for the even-mode both the signal carrying lines are at same potential, the electric fields also get terminated at the lateral ground as well as the ground plane below. Now the boundary condition used for computation with respect to the plane of symmetry pp' (refer to Fig. 3.10) changes to an electric wall $(x = 0)$ and $(x = g/2)$. For such a computation of the even-mode capacitance, we need to only consider the line to line separation as $s = 2*d_1$.

For computing odd-mode impedances, we assume electric boundaries at $x = 0$ and $x = c/2$. The odd-mode admittance Y_{odd} at the charge plane $y = y_0$ is given by:

$$Y+_{odd,n} = \varepsilon_0\left(\varepsilon_3 \coth\left(\beta'_n b_3\right)\right) \tag{3.19}$$

and

$$Y-_{odd,n} = \varepsilon_0\left(\varepsilon_3 \coth\left(\beta'_n b_2\right)\right) \tag{3.20}$$

where

$$\beta_n = \frac{n\pi}{c}, \ \beta'_n = \frac{n\pi}{c'}$$
$$c = 2(d + w) + s$$
$$c' \gg w$$
$$n = 2, 4, 6, \ldots \infty.$$

Here $Y+_{odd}$ and $Y-_{odd}$ are the admittances reflected on the charge plane due to the upper layer (ε_3, b_3) and the lower layer (ε_2, b_2), respectively. Using (3.19), (3.20), (2.43), one can calculate the odd-mode capacitance for the upper and lower layers, respectively. These two capacitances are then summed up. It is important to note that the wall-to-wall spacing for the upper layer is considered to be c' which is much larger than c for the lower layer. Thus, for computation of the upper layer capacitance one should replace c in (2.43) by c'.

The even-mode admittance Y_{even} at the charge plane $y = y_0$ is given by:

$$Y+_{even,n_1} = \varepsilon_0\left(\varepsilon_3 \coth\left(\beta'_{n_1} b_3\right)\right) \tag{3.21}$$

and

$$Y-_{even,n} = \varepsilon_0\left(\varepsilon_2 \coth(\gamma_n b_2)\right), \tag{3.22}$$

where

$$\gamma_n = \frac{n\pi}{g}, \ \beta'_{n_1} = \frac{n_1\pi}{c'}$$
$$c' \gg w$$
$$n_1 = 1, 3, 5, \ldots \infty \text{ and}$$
$$n = 2, 4, 6. \ldots \ldots \infty.$$

Again, the even-mode capacitances for the two vertical layers are computed independently. For the lower region (ε_2, b_2), the capacitance is computed using (3.21), (3.22) for even values of 'n' since the boundary condition has now changed, whereas for the upper layer capacitance is computed using (3.21), (3.22), (2.43). These two capacitances are then summed up.

Using (3.19–3.22) we can calculate the even- and odd-mode impedances. The voltage coupling coefficient C_v can be computed as:

$$C_v = \frac{Z_{even} - Z_{odd}}{Z_{even} + Z_{odd}}. \tag{3.23}$$

Also, the inductive and capacitive coupling coefficients can be computed as:

$$
\begin{aligned}
k_c &= \frac{Z_{even}\beta_{odd} - Z_{odd}\beta_{even}}{Z_{even}\beta_{odd} + Z_{odd}\beta_{even}} \\
k_l &= \frac{Z_{even}\beta_{even} - Z_{odd}\beta_{odd}}{Z_{even}\beta_{even} + Z_{odd}\beta_{odd}},
\end{aligned}
\tag{3.24}
$$

in the above discussions, we present systematic design steps that can aid in the computation of coupling coefficients in case of coupled interconnects lines. We now present the results obtained from the above model and compare them with 3D field simulations and measurements. While obtaining the simulated results we have considered 50 Ω terminations at all ends. We present two cases; in the first case $d \gg d_1$ while in the second case $d = d_1$. From the results presented in Tables 3.5 and 3.6, it is seen that the analytical results are within 2 dB accuracy of the results predicted by full-wave simulations. This difference may be attributed to the assumption of converting an inner ground track of finite dimension into an infinitesimally thin vertical ground plane with spacing between the two signal lines altered.

Based on the above discussion one can now compute even- and odd-mode admittances for any arbitrary-coupled interconnect lines. For a differential pair with intermediate shield insertion shown in Fig. 3.9b the admittance parameter can be given as:

$$Y = \varepsilon_0 \left[\coth\left(\beta_n' b_3\right) + \varepsilon_2 \coth\left(\beta_n'' b_2\right) \right]$$

where

$$
\begin{aligned}
\beta_n' &= \frac{n\pi}{c}, c \gg w \\
\beta_n'' &= \frac{n\pi}{c'}, c' = 2(d+w) + s
\end{aligned}
\tag{3.25}
$$

The model for differential pair with ground insertion is validated by comparing results with 3D field-solver as shown in Table 3.7 and shows good agreement. Table 3.8 gives the average computation time required to analyze the above-mentioned structures which is again much lesser than that required for running 3D simulations.

Table 3.5 Comparison between analytical and simulated results, ($w = 0.2$ mm, $b_2 = 0.508$ mm, $b_3 \gg b_2$, $d = 50$ mm, and $f = 5$ GHz)

Line to line spacing (s) in mm	Line to center ground spacing (d_1) in mm	$\varepsilon_2 = 2.2$			$\varepsilon_2 = 4.6$			$\varepsilon_2 = 9.9$			$\varepsilon_2 = 11.9$		
		Simulated (S_{31}) in dB (with center ground track)	Analytical (S_{31}) in dB (with center ground track)	Simulated (S_{31}) in dB (without center ground track)	Simulated (S_{31}) in dB (with center ground track)	Analytical (S_{31}) in dB (with center ground track)	Simulated (S_{31}) in dB (without center ground track)	Simulated (S_{31}) in dB (with center ground track)	Analytical (S_{31}) in dB (with center ground track)	Simulated (S_{31}) in dB (without center ground track)	Simulated (S_{31}) in dB (with center ground track)	Analytical (S_{31}) in dB (with center ground track)	Simulated (S_{31}) in dB (without center ground track)
0.1	0.01	−19.1	−19.42	−7.6	−22.43	−23.66	−7.8	−25.92	−27	−8	−27	−27.72	−8.3
0.1	0.02	−16.23	−17	−7.6	−19	−19.64	−7.8	−20.45	−21.6	−8	−21.3	−21.94	−8.3
0.15	0.02	−19.56	−20.36	−8.77	−23.67	−24.5	−9.06	−27.67	−28.3	−9.2	−28.12	−29	−9.3
0.15	0.05	−14.78	−15.51	−8.77	−16.86	−17.5	−9.06	−18	−18.94	−9.2	−18.48	−19.18	−9.3
0.2	0.05	−16	−17.43	−9.9	−19.1	−20	−10.2	−21.02	−21.94	−10.4	−21.76	−22.29	−10.5
0.2	0.07	−14.89	−15.61	−9.9	−16.92	−17.6	−10.2	−18.1	−19	−10.4	−18.36	−19.22	−10.5
0.25	0.07	−16.72	−17.1	−10.9	−18.74	−19.5	−11.31	−20.46	−21.25	−11.54	−20.58	−21.56	−11.6
0.25	0.1	−14.3	−15.2	−10.9	−16.46	−17	−11.31	−17.39	−18.25	−11.54	−17.37	−18.46	−11.6

Table 3.6 Comparison between analytical and simulated results, ($w = 0.2$ mm, $b_2 = 0.508$ mm, $b_3 \gg b_2$, $d = d_1$ mm, and $f = 5$ GHz)

Line to line spacing (s) in mm	Line to center ground spacing (d_1) in mm	$\varepsilon_2 = 2.2$		$\varepsilon_2 = 4.6$		$\varepsilon_2 = 9.9$		$\varepsilon_2 = 11.9$	
		Simulated (S_{31}) in dB (with center ground track)	Analytical (S_{31}) in dB (with center ground track)	Simulated (S_{31}) in dB (with center ground track)	Analytical (S_{31}) in dB (with center ground track)	(S_{31}) in dB (with center ground track)	Analytical (S_{31}) in dB (with center ground track)	Simulated (S_{31}) in dB (with center ground track)	Analytical (S_{31}) in dB (with center ground track)
0.1	0.01	−29.11	−30	−42.56	−43.97	−41	−42.7	−39.22	−40.54
0.1	0.02	−20.9	−21.26	−22.39	−24.41	−25.87	−26.76	−26.51	−27.74
0.15	0.02	−25.45	−26.27	−31.72	−32.8	−39.98	−41	−42.67	−43.41
0.15	0.05	−17.67	−17.55	−18.32	−19.52	−20.65	−21	−20.34	−21.19
0.2	0.05	−18.45	−19.6	−21.02	−22.13	−23.21	−24	−23.81	−24.37
0.2	0.07	−16.78	−17.25	−18.19	−19.21	−19.81	−20.54	−19.9	−20.76
0.25	0.07	−17.52	−18.75	−20.31	−21	−21.41	−22.72	−22.11	−23
0.25	0.1	−15.09	−16.47	−17.02	−18.22	−18.36	−19.4	−18.02	−19.67

Table 3.7 Crosstalk comparisons between model and field simulator, ($w = s = d$, $\varepsilon_2 = 4.6$, $f = 10$ GHz)

Line to line spacing (s) in mm	Simulated (S_{31}) in dB	Analytical (S_{31})in dB	Analytical (S_{31}) in dB (for $d \gg 2w + s$)
0.25	−14	−13.89	−11
0.5	−18.62	−18.1	−17
0.75	−23.16	−22.87	−20.9
1	−26.6	−25.9	−24.1
1.25	−29.32	−28.78	−27.4
1.5	−31.53	−30.81	−29.6

Table 3.8 Parameter extraction time using our proposed algorithm

Interconnect structure	Extraction time (in seconds)
Coupled interconnect lines with intermediate shield insertion	43
Differential pair with shield insertion	46

3.3 Concluding Remarks

The case studies presented in this chapter highlight the applicability of unified approach in analyzing a variety of interconnect structures. The results can be an important design parameter and can be implicitly linked to overall signal integrity. The model once programed can give results fairly, quickly and accurately. It should be noted that the unified approach-based compact models are quasistatic and therefore limited to an upper frequency. Beyond this frequency, dispersive effects are observed and suitable extension to these models can be built to take the dispersion into account. The modeling approach is very effective for CAD-based application and particularly suited for planar interconnect structures. The models do not use any inherent assumptions. The assumption of trial function $f(x)$ is based on maximizing the capacitance that can give the closest value to exact solutions and is based on the charge distribution in the interconnect cross-section.

In the above discussion an important insight is also obtained by applying the approach to both single and coupled lines. We observe that by inserting a ground trace between two parallel lines in proximity crosstalk is reduced, yet such proximity of ground trace to individual lines will change the characteristic impedance. Therefore, such an exercise is an iterative process; and that is not intuitive if the designer applies numerical solutions straightway. We may also consider modifications in ground aperture in different ways to derive unusual results. As for example, a single, straight and uniform signal trace can show multiple impedance steps (thereby impedance transformation) by modifying the ground around and below the line selectively (at different locations with respect to the line). Similarly, using ground aperture size judiciously for an edge-coupled microstrip coupler, we will be able to tighten the coupling as compared to the one possible in conventional design. Microwave engineers will love that! In the next chapter an overview of the measurement techniques and simulation methodologies is presented.

References

1. B. Bhat, S.K. Koul, Unified approach to solve a class of strip and microstrip-like transmission lines. IEEE Trans. Microw. Theory Tech. **82**(5), 679–686 (1982)
2. B. Bhat, S.K. Koul, *Stripline-like Transmission Lines for Microwave Integrated Circuits* (Wiley, New york, 1989)
3. R.E. Collin, *Field Theory of Guided Waves* (McGraw-Hill, New York, 1960)
4. H.E. Green, The numerical solution of some important transmission line problems. IEEE Trans. Microw. Theory Tech. **13**(5), 676–692 (1965)
5. R. Crampagne, M. Ahmadpanah, J.L. Guiraud, A simple method for determining the green's function for a large class of MIC lines having multilayered dielectric substrates. IEEE Trans. Microw. Theory Tech. **26**(2), 82–87 (1978)
6. C. Paul, *Analysis of Multiconductor Transmission Lines*, 2nd edn. (Wiley, Hoboken, 2007)
7. T.C. Edwards, M.B. Steer, *Foundations of Interconnect and Microstrip Design*, 3rd edn. (Wiley, New York, 2001)
8. J. Fan et al., Signal integrity design for high-speed digital circuits: progress and directions. IEEE Trans. Electromagn. Compat. **52**(2), 392–400 (2010)
9. E.-P. Li et al., Progress review of electromagnetic compatibility analysis technologies for packages, printed circuit boards, and novel interconnects. IEEE Trans. Electromagn. Compat. **52**(2), 248–265 (2010)
10. R. Sharma, T. Chakravarty, A.B. Bhattacharyya, Analytical modeling of microstrip-like interconnects in presence of ground plane aperture. IET Microw. Antennas Propag. **3**(1), 14–22 (2009)
11. R. Sharma, T. Chakravarty, A.B. Bhattacharyya, Analytical model for optimum signal integrity in pcb interconnects using ground tracks. IEEE Trans. Electromagn. Compat. **51**(1), 67–77 (2009)
12. T. Zhang, S.S. Sapatnekar, Simultaneous shield and buffer insertion for crosstalk noise reduction in global routing. IEEE Trans. VLSI Syst. **15**(6), 624–636 (2007)
13. J. Zhang, E. G. Friedman, Crosstalk modeling for coupled RLC interconnects with application to shield insertion. IEEE Trans. VLSI Syst. **14**(6), 641–646 (2006)
14. R. Simmons, *Coplanar Waveguide Circuits, Components, and Systems*, 1st edn. (Wiley-IEEE press, New York, 2001)
15. S.M. Musa, M.N.O. Sadiku, Capacitance and inductance matrices for multistrip lines. in *Proceedings of the COMSOL Conference* (2007)
16. N.D. Arora, Challenges of modeling VLSI interconnects in the DSM Era. in *Proceedings of International Conference on Modeling and Simulation of Microsystems* (2002), pp. 645–648
17. Y. Cao et al., Impact of on-chip interconnect frequency-dependent R(f)L(f) on digital and rf design. IEEE Trans. VLSI Syst. **13**(1), 158–162 (2005)
18. Y.I. Ismail, E.G. Friedman, J.L. Neves, Equivalent elmore delay for RLC trees. IEEE Trans. CAD Integr. Circuits Syst. **19**(1), 83–97 (2000)
19. Y.I. Ismail, E.G. Friedman, J.L. Neves, Equivalent elmore's delay for RLC trees. in *Proceedings of Design Automation Conference* (1999), pp. 715–720
20. R. Sharma, T. Chakravarty, K. Choi, Fast and efficient extraction algorithm for high-speed interconnects with arbitrary boundaries. J. Supercomput. (in press). doi:10.1007/s11227-011-0713-2

21. I. Novak, B. Eged, L. Hatvani, Measurement and simulation of crosstalk reduction by discrete discontinuities along coupled PCB traces. IEEE Trans. Instrum. Meas. **43**(2), 170–175 (1994)
22. S. K. Kim, C. C. Liu, L. Xue, S. Tiwari, Crosstalk attenuation with ground plane structures in three-dimensionally integrated mixed signal systems, IEEE MTT-S Symposium (2005), pp. 2155–2159
23. J.C. Coetzee, J. Joubert, Full-wave characterization of the crosstalk reduction effect of an additional grounded track introduced between two printed circuit tracks. IEEE Trans. Circuits Syst. **43**(7), 553–558 (1996)
24. H. Ozaki, J. Ishii, Synthesis of a class of stripline filters. IRE Trans. Circuit Theory **5**(2), 104–109 (1958)
25. M.K. Krage, G.I. Haddad, Characteristics of coupled microstrip transmission lines-I: Coupled-mode formulation of inhomogeneous lines, II: Evaluation of coupled line parameters. IEEE Trans. Microw. Theory Tech. **18**(4), 217–222 (1970)
26. R. Pregla, Calculation of the Distributed Capacitance and Phase Velocities in Coupled Microstrip Lines by Conformal Mapping Techniques. AEÜ **26**, 470–474 (1972)
27. T.G. Bryant, J.A. Weiss, Parameters of microstrip transmission lines and of coupled pairs of microstrip lines. IEEE Trans. Microw. Theory Tech. **16**(12), 1021–1027 (1968)
28. L. Young, H. Sobol, *Advances in Microwaves* (Academic Press, New York, 1966)
29. H.G. Bergandt, R. Pregla, Calculation of even- and odd-mode capacitances parameters for coupled microstrips. AEÜ **26**, 153–158 (1972)

Chapter 4
Measurement Technique and Simulation Methodology

Modern technology trend is pushing the performance requirements for chip to chip interconnects in the microwave frequency band (typically above 10–12 GHz). Now, we have a paradigm where mixed signal and complex PCB layouts are required to operate optimally even up to X-band (8–12 GHz). For such cases where the interconnect dimensions are of the order of wavelength, transmission line effects become the dominant factor. We then benchmark the performance criteria by measuring reflection, crosstalk, propagation loss, delay and dispersion. Operating at microwave frequency range also necessitates that the interconnect design process takes into consideration the characteristic impedance of the line as well as source/ load impedance matching. A working design is helped manifolds by accurate modeling approach; some of these are described in previous chapters. We may model interconnect as lossless transmission line. But in real applications, at microwave frequency range, such transmission lines will display frequency dependent losses. This will result in signal integrity issues like deterministic jitter etc. Thus, it is imperative that we validate the models by carrying out accurate measurements. In the modeling approach, we have presented analytical method of determining the various performance criteria. But, such analytical approach cannot be stated as 'exact'. To obviate resource intensive physical iteration, it is always preferred to do electromagnetic simulation of the structure. Using analytical techniques in conjunction with EM simulation reduces the time-to-market of the device. In this chapter, we present an overview of the measurement techniques as well as EM simulations for the purpose of evaluating high-speed interconnect designs.

4.1 Measuring Characteristic Impedance

The question props up again; what do we want to measure and how? In the previous chapters, we have discussed the modeling aspects of the high speed interconnect. We make an attempt to analyze the performance of the interconnect

R. Sharma and T. Chakravarty, *Compact Models and Measurement Techniques*
for High-Speed Interconnects, SpringerBriefs in Electrical and Computer Engineering,
DOI: 10.1007/978-1-4614-1071-3_4, © The Author(s) 2012

Fig. 4.1 Schematic of
measurement setup for
characteristic impedance

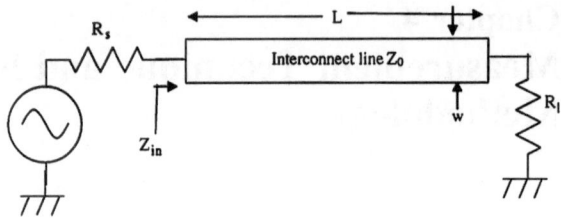

line in terms of width 'w', height 'b' of the line over ground, proximity of the line
to other interconnects etc. For all such cases, the constraints are many, such as
dielectric constant, source impedance and load impedance among others. The
characteristic impedance Z_0 is the vital parameter to evaluate (forming the basis of
a good interconnect model), and that is obtained in terms of the ratio, w/b for a
given substrate. The Z_0 value determines the extent of reflection which is measured
as reflection coefficient and given in (4.1).

$$\rho = \frac{V_{reflected}}{V_{incident}} = \frac{Z_{load} - Z_0}{Z_{load} + Z_0} \tag{4.1}$$

In Fig. 4.1, the schematic of measurement setup for characteristic impedance is
shown.

As shown here, the interconnect line is a two port network (interchangeable
ports) with port 1 considered as input port and port 2 as output port. To obtain the
value of the unknown parameter Z_0 through measurement, we must know the
source and load impedance. We inject a high frequency signal at port 1 and
measure reflection coefficient. However, the length of the interconnect line needs
to be accounted for, unless the electrical length of the line is very short. The length
effect is considered using the following relationship:

$$Z_{in} = Z_0 \frac{Z_L + jZ_0 \tan \beta L}{Z_0 + jZ_L \tan \beta L} \tag{4.2}$$

here β is the propagation constant and Z_L is the load impedance.

For two selections of load impedance, namely $Z_L = 0$ (Short Circuit; SC) and
$Z_L = \infty$ (Open Circuit; OC), short and open circuited input impedances [1] can be
obtained from (4.2) as given below:

$$Z_{inSC} = jZ_0 \tan \beta L \tag{4.3}$$

$$Z_{inOC} = -jZ_0 \cot \beta L \tag{4.4}$$

The characteristic impedance Z_0 is now given by:

$$Z_0 = \sqrt{Z_{inSC} Z_{inOC}} \tag{4.5}$$

In (4.5), the measured parameters are 'short-circuit' impedance and 'open-circuit' impedance respectively while characteristic impedance value is deduced. For most accurate and repeatable results, the line length should be as short as possible. Also, the measurement with OC load requires care as the effects of fringing fields will alter the effective length of the interconnect line. However, we can avoid 'OC' by making 'SC' measurements of two (or more) interconnect lines of different lengths (except quarter wave length line) and using (4.3).

4.2 Frequency Domain Measurement Technique

To obtain the Z_0 value, we need to measure complex impedances. Complex impedance value is frequency dependent; thus, the frequency of measurement is a critical component for such measurements. Frequency domain measurement is done by Vector Network Analyzer (VNA), using swept frequency measurement type [2]. The two ports in VNA can be utilized independently or together (for through measurements). VNA is used to measure reflection coefficient, thereby input/output impedance and transmission loss as a function of frequency. VNA measurement is generally applicable for measurement on passive devices or the ones that do not translate the frequency like mixers etc. Vector measurement signifies that the VNA measures both amplitude and phase of the signal. The measurements are done for scattering parameters (S-parameters). For a two port network, the measured parameters are $S_{ij} = \{S_{11}, S_{12}, S_{21}, S_{22}\}$ in dB, where 'i' denotes output port and 'j' denotes input port. However, for Z_0 measurement, only one port is required, S_{11} or S_{22}. A typical S_{11} measurement is shown in Fig. 4.2a where X-axis is frequency (GHz) and Y-axis is amplitude (dB). We need to define the 'START' and 'STOP' frequency. It is to be noted that the characteristic impedance is independent of frequency of operation. Therefore, we can make the measurement at lower frequencies as compared to the desired frequency. This is called quasi-static measurement. When one port of the VNA is connected to the interconnect line (using a coaxial to microstrip transition), we obtain input impedance in the form of Smith chart giving real and imaginary part of the said impedance. A representative sample measurement is shown in Fig. 4.2b.

A set of measurements at the maximum desired frequency is also required; these are likely to show measurement deviation due to 'dispersion' effect. A typical interconnect line is likely to have two different media at the interface, e.g. air and dielectric material. Here, the group velocity changes with frequency in a non-linear manner. This effect is reflected on effective dielectric constant as well as characteristic impedance. On thin, low relative dielectric constant and low dielectric loss substrates, the dispersion effect becomes significant only beyond approximately 10–12 GHz [3]. However, on standard low cost FR-4 substrates this dispersive effect can be seen at even lower microwave frequency range. Due to dispersion effect, effective dielectric constant increases with frequency till it

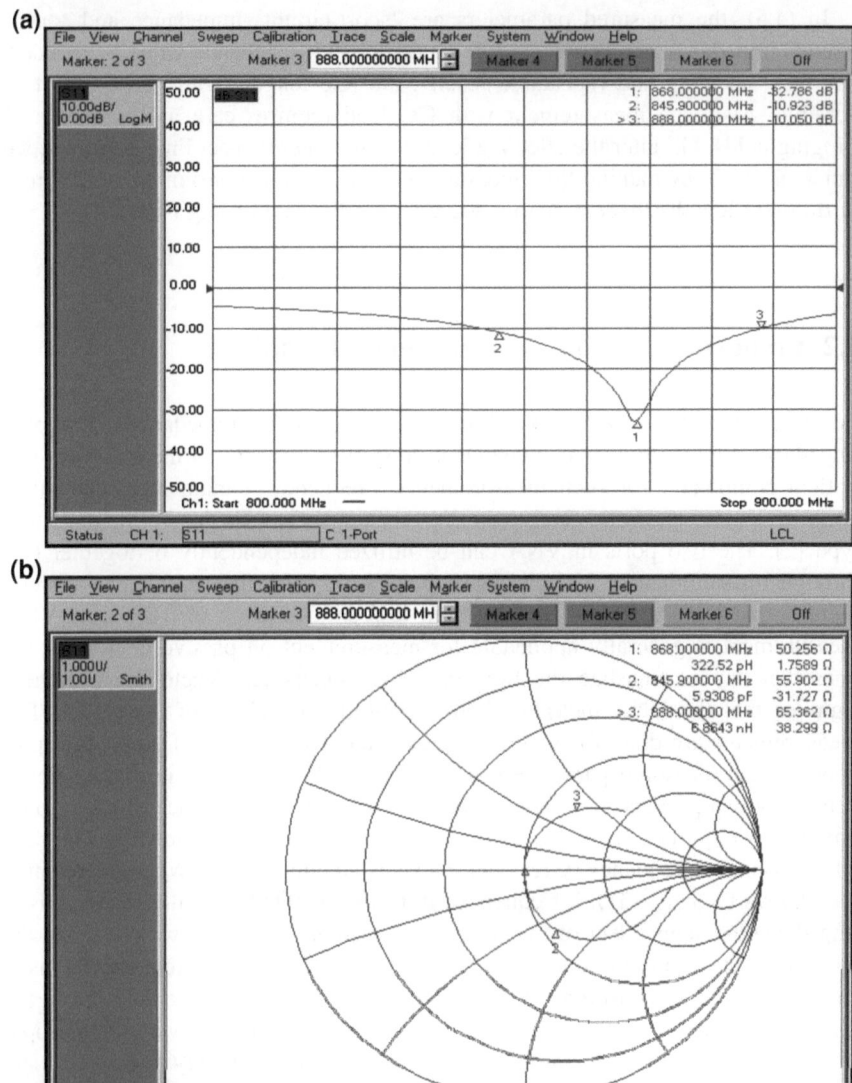

Fig. 4.2 Typical VNA measurement screen **a** S_{11} (dB), **b** Smith chart

reaches an asymptotic value. Dispersion is quantified by measuring "Group Delay" for S_{21}, which happens to be the transmission coefficient. For non-dispersive interconnect designs (like stripline), the group delay is constant while for conventional microstrip type interconnects, group delay changes with frequency. Group delay is given as approximately 1 nano second per foot multiplied by square

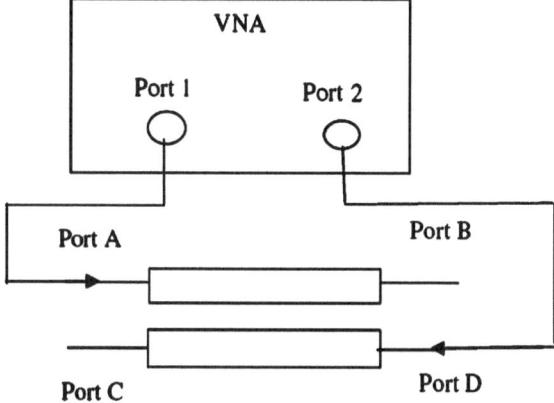

Fig. 4.3 Schematic of measurement setup for measuring crosstalk (unused ports are terminated)

root of effective dielectric constant (which is much less than dielectric constant for narrow lines). As effective dielectric constant increases, Z_0 begins decreasing.

From the above discussion, we can conclude that for accurate VNA-based measurement of Z_0 of an interconnect line; the following steps are required to be taken.

• Build a short length of the line on the substrate selected and bond RF connectors on both the ports.
• Calibrate the VNA using precision reference impedance for de-embedding the effects of coax to microstrip transitions.
• Carry out 1 port reflection coefficient measurement at lower frequency (VHF/UHF) to accurately determine the quasi-static value of Z_0.
• Repeat the measurements up to the maximum desired frequency and plot measured Z_0 with respect to frequency. The plot will identify a corner frequency for the deviation of Z_0.
• Deduce dispersion effects like change in effective dielectric constant, loss tangent etc. and conclude if the selected substrate can be used for the desired high-speed application.
• Rework on the analytical models based on measured data.

VNA can also be used to measure crosstalk or mutual coupling. Figure 4.3 displays the basic measurement set up for measuring crosstalk in coupled interconnect lines.

Here, we have two interconnect lines in close proximity. We assume that the two lines are driven by separate sources and are driving separate loads. Thus, we can treat the structure as a four port network. To measure the crosstalk we use S_{21} measurement. We feed port 'A' of the coupled structure from port 1 of VNA and measure crosstalk on port 'C' or 'D'. Since by nature microstrip lines are backward coupler, it is likely that port 'C' will display higher value of coupling (called coupling factor) than port 'D' will show (called isolation factor). In this configuration port 'B' should be terminated with the desired load impedance.

A word of caution though; VNA measurement assumes 50 Ω for the Device-Under-Test (DUT), while chips are likely to offer complex impedance to the interconnect line. Therefore, it becomes imperative to de-embed [4] the effects of external connectors used in the measurement set up. At microwave frequencies we normally use N- or SMA-type coaxial connectors. Such de-embedding process requires careful modeling and validation of these fixtures. Thereafter, the simulated response is integrated to the VNA's existing error model. Many a times, it is sensible to avoid the de-embedding process by using direct calibration. Direct calibration requires three high precision impedance standards namely short, open and load. In addition 'thru' connection is used to calibrate transmission loss. We can also consider using alternate calibration process called Thru-Reflect-Line (TRL) [5]. For such cases, the only precision impedance reference required is a transmission line section. TRL calibration occurs in three steps: in the first step, the test fixtures for the two ports of VNA are connected together (THRU) and all four S-parameters are measured. In the second step, a high reflection coefficient device is connected to both the ports (REFLECT) and measurements are repeated. In the final step, a short section of microstrip line is inserted between the two fixtures. The characteristic impedance of the line becomes the absolute impedance reference.

4.3 Time Domain Measurement Technique

A swept frequency measurement using VNA offers high accuracy due to the fact that the network analyzers have very high dynamic range (>100 dB). But such high precision comes at a hefty price. Moreover, signal integrity tests do not usually require such high dynamic range [6, 7]. Digital designers also have difficulty in appreciating a detailed calibration process for a VNA. A very useful alternative to VNA is time domain measurements using Time Domain Reflectometry (TDR). TDR is a sampling oscilloscope with a fast rise time of the order of less than 25ps. This is equivalent to very high frequency bandwidth extending to few GHz! TDR is known to be used for 'fault location' of a long transmission line but a lot of useful information can be deduced by observing the reflections happening on the transmission line at different time frames. TDR sends a 'step' stimulus to the device and reflections are measured to characterize lossy transmission lines, rise time degradation, reflection coefficient, skin effect etc. We can use TDR for in situ measurements provided we account for the degradation in rise time due to the probes and fixtures using (4.6) given below:

$$t_{measured} = \sqrt{t_{TDR}^2 + 2[\frac{0.35}{f_{3dB}}]^2} \qquad (4.6)$$

Fig. 4.4 Pulse types for TDR/TDT measurement **a** Step, **b** impulse

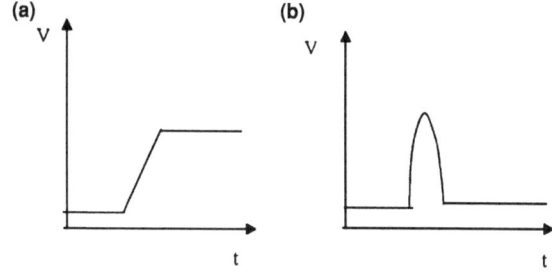

here t_{TDR} is the rise time measured without the probes and $f_{3\ dB}$ is the 3 dB bandwidth of the cable and probes, combined. The degradation is minimized by providing for both signal and ground connections with minimum spacing. For a single uniform interconnect line, assuming that there are no discontinuities in the line, multiple reflections in a TDR measurement are avoided thereby simplifying the use of TDR.

Along with TDR, we would also require Time Domain Transmission (TDT) measurements. In the frequency domain, such measurement pair relates to S_{11} and S_{21} respectively. Referring to Fig. 4.1, we need to replace a swept synthesizer (sine wave) by pulse generator. The impressed signal can be a step or an impulse as shown below in Fig. 4.4 and explained in [8].

The frequency bandwidth of a step function is $1/t_r$ even though 75% of that is usually considered a useful spectrum [7]. Some manufacturer of TDR equipment like Picosecond Pulse Lab (PPSL) has models that produce steps of the order of 5ps step rise time and 15ps impulse duration. Thus, we can theoretically measure very high speed indeed! We consider that the impulse response of the interconnect line be $h\ (t)$, the input signal be $V_{in}\ (t)$ and the output signal be $V_{out}\ (t)$. Therefore the output voltage V_{out} is given by:

$$V_{out}(t) = v_{in}(t) \otimes h(t) \ldots \text{in time domain}$$
$$V_{out}(f) = V_{in}(f)H(f) \ldots \text{in frequency domain} \tag{4.7}$$

For TDR measurement, the oscilloscope measurement is done at the input port only. As the impressed pulse propagates through the transmission line, for every impedance discontinuities in the line, a portion of the incident energy is reflected back. The time taken by the reflected signal (echo) with respect to the rising edge of the impressed pulse is measured, divided by two two-way travel) and the wave shape is analyzed to identify the location and type of discontinuity as shown in Fig. 4.5. In Fig. 4.5, different waveforms are shown for different values of characteristic impedance as well as complex terminations. Usually, the interconnect line feeds another chip which is most likely to be treated as a termination of type RC in series. For a resonant circuit type mismatch, damped oscillations are observed.

By measuring the time constant of the amplitude varying waveform, the values of L and C can be obtained as

$$L = Z_0 \cdot \tau, \ C = {}^\tau\!/\!_{Z_0} \tag{4.8}$$

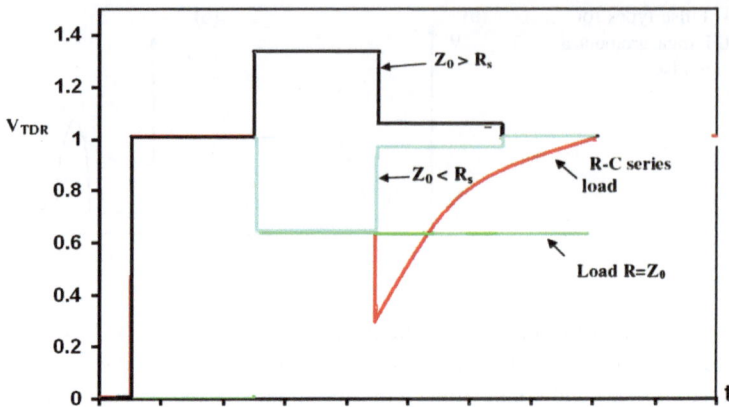

Fig. 4.5 TDR measurement for different terminations

From Fig. 4.5 and application of (4.8), we observe that it is possible to determine Z_0 by knowing the reactive termination accurately under the assumption that the line does not contain any other impedance step (essentially a straight line nearly matched to the source impedance). By measuring the time duration (of the discontinuity of load), it is possible to measure the effective dielectric constant also. For example, if we terminate the line with inductor/capacitor (refer Fig. 4.5) then by measuring the time 't' till the termination is denoted by the sharp change in amplitude, we can calculate the electrical distance as given below:

$$t = 2 \cdot l / V_p \tag{4.9}$$

where

$$V_p = c / \sqrt{\varepsilon_{eff}} \tag{4.10}$$

In (4.9) and (4.10), we have measured the time in oscilloscope, known the physical length (l) with precision; therefore we can deduce propagation velocity in the material.

From the above discussions, it is observed that TDR/TDT measurement is valuable for being able to differentiate complex impedance loads. Yet, it will not be entirely incorrect to state that small step degradation (due to changes in impedance over time) is difficult to quantify using TDR. For such, we can use another complementary method called Frequency Domain Reflectometry (FDR).

4.4 Frequency Domain Reflectometry Technique

FDR is more useful when measuring small performance changes [9]. Unlike the VNA method; with FDR technique, it is possible to exactly locate the position of impedance discontinuity. TDR is known to do this by application of pulses (with

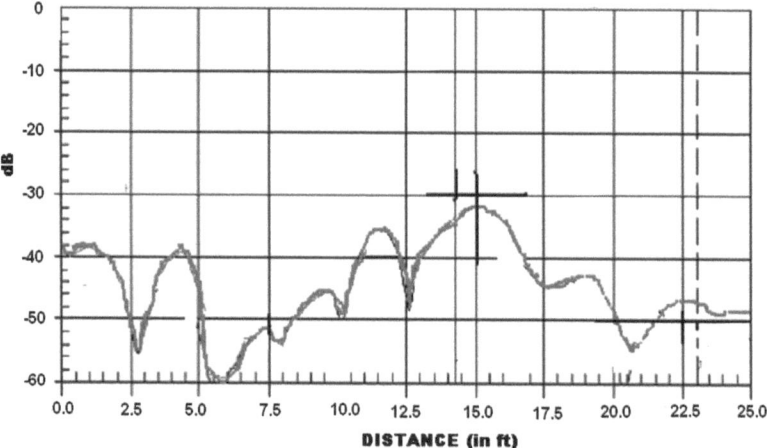

Fig. 4.6 FDR representation (return loss versus distance)

DC content). FDR does the same by using swept high frequency signal. Similar to the set up shown in Fig. 4.1, in FDR measurement, we apply frequency sweep at the input of the transmission line terminated by a load. The reflected signal is measured in the frequency domain itself. The signal processing method of Inverse Fast Fourier Transform (IFFT) is then applied on the frequency domain reflection coefficient in order to convert it to time domain. The time domain data is then plotted with respect to the distance (of propagation of wave) from the point of launch to give us the impedance profile with distance as shown in Fig. 4.6. Return loss is the vector sum of the reflected signals measured at the input port at one frequency.

By doing IFFT, we are extracting individual reflection components occurring from different sections of the transmission line. This becomes very useful for interconnect lines with varying impedance profile like sharp bends, tapers, pads on the line etc. For example, we refer to Fig. 4.7, where two arbitrary lines are drawn with partial coupling and right angle bends.

Consider that there are two interconnect lines 'A' and 'B' having two input ports; 1 and 3, (on the U1 side). If we carry out a VNA measurement at the input ports, we will see a composite return loss S_{11} for each line along with S_{31} to represent mutual coupling. Such measurement is relevant at the reference plane; that is the input ports. In FDR technique, the return loss is converted to time domain, thereby displaying reflection coefficient with respect to distance as measured from the input port. We will then see changes in impedance profile at the discontinuous points like coupling region and bends. Such measurements are extremely valuable for modeling the interconnect lines.

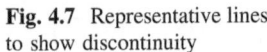

Fig. 4.7 Representative lines
to show discontinuity

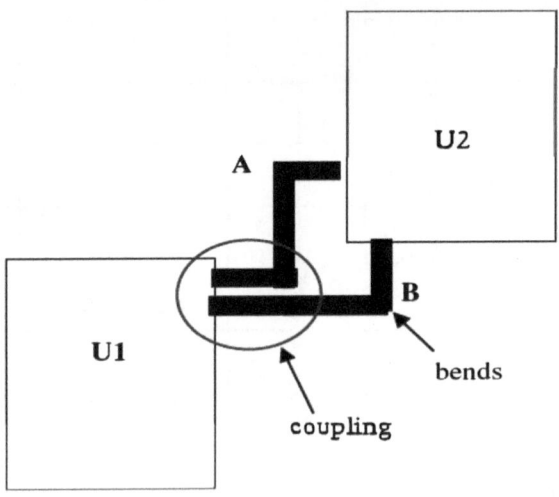

4.5 Simulation of Interconnect Lines

We have discussed the methods of building models for high speed interconnects; in particular, considered the effects of ground plane apertures and ground track insertions. Such models are at best approximate, with accuracy depending on the electrical size. An exact solution will be available if we carry out full-wave electromagnetic simulation of such lines [10]. Yet, the attractiveness of having an approximate model for first level synthesis and analysis is immense. It saves time, saves resources and gives physical insights. The natural progression from model based design leads to measurement as the final step. However, if modeling errors lead to repeated measurements and increased iteration, much of the above stated accrued benefits due to modeling are lost. Such a situation is averted by an intermediate stage of numerical simulation of the structure. A representation of the thought process is given in Fig. 4.8.

Figure 4.8 displays the process of executing design optimization in two sequential steps. The initial design is put in place quickly using analytical techniques. Currently, we have host of literature giving closed-form expressions for different type of distributed lines, gaps, bends and impedance steps [3]. These are also available as building blocks in circuit simulation tools like Agilent ADS [11]. More complex geometries can also be analyzed using the method outlined in this book. Circuit simulation tools enable us to iterate, if necessary till a pre-assigned output parameter value is obtained with some defined accuracy and tolerance. In the next step, this geometry is imported into an EM simulator based on Finite Difference Time Domain (FDTD), Finite Element Method (FEM) or Method of Moments (MoM). There are a number of commercial solvers with proven ability like CST Microwave Studio® [12], HFSS™ [13], IE3D [14], Sonnet [15] etc. In this stepped process, only a few runs are required to finalize the design followed by

Analytical model based   **Electromagnetic tool based**

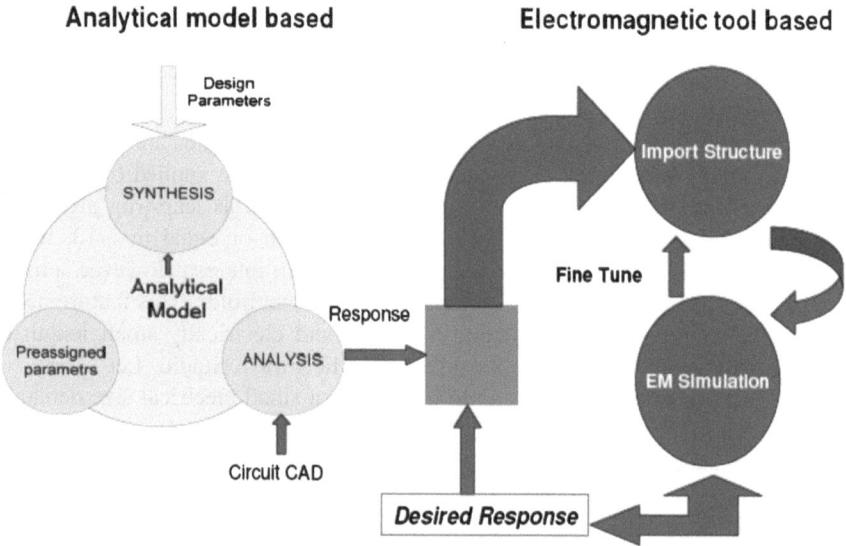

Fig. 4.8 Design and optimization cycle

fabrication and measurements. Simulation can also be used to fine tune the analytical models but for that it is important to differentiate the differences in the inherent assumptions. For example, a track on small sized PCB will give marginally different results for characteristic impedance (in EM solver) as compared to the closed-form expressions for standard microstrip (or similar) lines.

An excellent reference literature by Gustrau et al. [10] has detailed out the intricacies of full-wave simulation method using the above stated EM software tools.

4.6 Numerical Modeling of Interconnect

We consider a basic structure of a conducting trace on a PCB where the bottom layer is referred to as ground. There are two numerical methods to model this basic structure. We can consider the trace as a surface and apply *surface integral method* or we can think of the structure as of a finite dimension inside 3D computational domains and apply *volume-based method*. In the first case, the material interface is segmented whereas for the later a discretization of the truncated 3D space around the object is done. In addition, the numerical solution is further classified as frequency domain or time domain. FDTD is an example of volume-based time domain method, FEM is volume-based frequency domain method and MoM is surface integral type frequency domain method.

In FDTD method, the electric and magnetic field components, as obtained from Maxwell's partial differential equation in time domain, are discretized in both space and time. The 3D computational domain is divided into orthogonal cells (like cuboids) which are electrically small and homogenous in nature. The electric field components lie on the edges of the cell and magnetic fields are on the center of the surfaces. This concept of mutual circulation was first applied by Yee [16]. The discretization in time leads to time-stepping known as leap-frog algorithm. For numerical stability, each cell should be less than or equal to $\lambda/10$, where wavelength (λ) is calculated at the highest frequency of interest. However, a much smaller cell size will increase the computation time manifolds. This feature needs to be kept in mind while simulating a narrow and electrically small length of interconnects using FDTD software like CST Microwave Studio. Let us say we are interested in evaluating only up to 6 GHz but a small electrical size demands that the minimum wavelength (λ_{min}) or maximum frequency (f_{max}) is chosen in such a way that at least a few numbers of cells are formed to cover both the width and length of the line.

In FEM method, we utilize variational calculus and the starting point is the time-harmonic Helmholtz equation. In variational method, we use energy type expressions like stored electric/magnetic energy or power flow. We define this as functional F. This functional F becomes minimal (stationary) for the unknown solution. Thus

$$\partial F = 0 \quad \text{and} \quad \partial^2 F < 0$$

In order to solve the variational expression numerically, we divide the volume in subdomains called "Finite Elements". For commercial EM software, the subdomain space is typically tetrahedrons. Essentially we create a mesh of the entire computational domain by these finite elements so that the whole structure is filled up. Within each element, we define the potential as a linear polynomial and write the unknown potential in terms of shape function. From these, we build the element matrix equation for a single finite element. In the next step, all the element matrices are compiled into a global matrix equation that contains all unknown potential values. Volume-based techniques require accurate definition of boundary conditions at the outer surface of the computational domain. We can use boundary conditions such as a perfect absorbing layer, electric wall, magnetic wall etc. FEM software permits boundary condition to be applied on curved surfaces. Some FEM tools use adaptive meshing technique where the mesh is optimized in a sequence of simulation runs. In this manner, we can begin with a coarse mesh and refine it automatically with each run. This adaptive process gives great flexibility.

We may want to simulate our interconnect model over a large bandwidth as well as the highest frequency of operation. For wide bandwidth simulations, time domain methods like FDTD are best as a single run provides the entire frequency response. However, FEM may be used to evaluate at specific zone of high frequencies to check if any resonance like effect is seen. This is more useful for reactive loading of the lines.

4.7 Concluding Remarks

In this chapter, we have presented, in brief, fundamental measurement techniques useful for evaluating the performance of interconnects. We have stressed upon the need for understanding the differences in setup as well as inherent assumptions while measuring. VNAs are generally used by microwave engineers to carry out accurate and sensitive measurements. However, we do not always need the high dynamic range available with VNA. Suitable time-domain techniques like TDR/TDT are equally likely to offer accurate results. Nowadays, however, FDR technique is preferred since it offers the advantage of differentiating even small changes in performance. Accurate measurement is a vital tool to validate and refine the models. Same can be stated for EM simulations. With proper understanding of the assumptions, it is definitely possible to get exact results even for complex layouts.

References

1. A. Das, S.K. Das, *Microwave Engineering* (McGraw-Hill Higher Education, New York, 2006)
2. Agilent Technologies, Understanding the fundamental principles of vector network analysis. Application note: AN-1287-1, Palo Alto (2000)
3. K.C. Gupta, R. Garg, C. Rakesh, *Computer-Aided Design of Microwave Circuits* (Artech House, Norwell, 1981)
4. R.F. Bauer, P. Penfield, De-embedding and unterminating. IEEE Trans. Microw. Theory Tech. **22**(3), 282–288 (1974)
5. S.A. Warternberg, *RF Measurements of Dies and Packages* (Artech House, Norwood, 2002)
6. TDA Systems, TDR and VNA measurement primer. Application Note, Lake Oswego, OR (2004)
7. TDA systems, Complete methodology for signal integrity analysis of gigabyte interconnects, Application note, Lake Oswego, OR (2003)
8. J. R. Andrews, Time domain reflectometry (TDR) and Time domain transmission (TDT) measurement fundamentals, AN-15. (Picosecond Pulse Labs, Boulder, 2004), http://www.picosecond.com
9. Anritsu, *Distance to Fault-Application Note* (Morgan Hill, 2011), http://www.anritsu.com
10. F. Gustrau, D. Manteuffel, *EM Modeling of Antennas and RF Components for Wireless Communication Systems* (Springer, New York, 2006)
11. Agilent Corporation, Advanced design system (ADS)—user's guide, Palo Alto (2011), http://www.agilent.com/find/eesof-ads2011
12. Microwave Studio-User's Guide, Darmstadt (2011), http://www.cst.com/Content/Products/MWS/Overview.aspx
13. Ansys Incorporated, ANSYS HFSS-product overview, Pennsylvania (2011), http://www.ansoft.com/products/hf/hfss
14. Mentor Graphics, Electromagnetic design (EM) and simulation software, Oregon (2011), http://www.mentor.com/electromagnetic-simulation
15. Sonnet Software, EM analysis and simulation, Syracuse, NY (2011), http://www.sonnetsoftware.com
16. K.S. Yee, Numerical solution of initial boundary value problems involving Maxwell's equations in isotropic media. IEEE Trans. Antennas Propag. **14**(3), 302–307 (1966)